URBAN AGRICULTURE:

GROWING HEALTHY, SUSTAINABLE PLACES

Kimberley Hodgson, AICP, Marcia Caton Campbell,
AND Martin Bailkey

TABLE OF CONTENTS

CHAPTER 1

Introduction

 The American Planning Association initiated this report to introduce practicing planners and local government representatives to the concept of urban agriculture, its different forms of practice across North America, and its connections to other social, economic, and environmental goals. The creation of this report coincides with the steady rise in popularity of urban agriculture in the United States and Canada, as evidenced by coverage in the popular press, its increasingly central place within the growing local food movement, and the increase in interest in planning cities to foster both healthier residents and more sustainable communities.

Urban agriculture entails the production of food for personal consumption, education, donation, or sale and includes associated physical and organizational infrastructure, policies, and programs within urban, suburban, and rural built environments. From community and school gardens in small rural towns and commercial farms in first-ring suburbs to rooftop gardens and bee-keeping operations in built-out cities, urban agriculture exists in multiple forms and for multiple purposes.

While it is a small component of the larger community-based food system, urban agriculture is important to the overall health and resilience of communities and regions, and as a practice it is expected to increase over the next decade. Therefore, urban agriculture has implications for urban planning as regulated by local and regional governments and planning agencies. These implications need analysis and clarification, since urban agriculture falls somewhat outside the range of traditional land-use designations (e.g., is a commercial urban farm as a land use most similar to a rural farm, a commercial enterprise, or a public park?). There are also emerging connections between urban agriculture and the redevelopment of urban brownfields in residential and industrial areas (see, e.g., Kaufman and Bailkey 2004), as well as the more extensive and more productive use of lawns and green space. Urban agriculture has been found to influence the value of neighboring real estate and thus has implications for land use beyond the boundaries of a particular agricultural site (Voicu and Been 2008).

> Urban agriculture has implications for urban planning as regulated by local and regional governments and planning agencies.

Kimberley Hodgson

Along with its connections to land-use planning, urban agriculture can contribute significantly to the development of social connections, capacity building, and community empowerment in urban neighborhoods, most commonly through community gardening (Hynes 1996; Johnson 2010). In addition, it offers links to community development practice as a viable means of creating jobs, training youth, supplementing food budgets, and generating modest levels of revenue for urban farmers who sell their products. Urban agriculture also has much to offer community health planners as a health-promoting activity but also as a mechanism to connect urban and suburban producers of fruits and vegetables with urban consumers. When combined with other efforts to improve access to healthy, affordable food (such as healthy-corner-store programs and supermarket-financing initiatives), urban agriculture can become a valuable tool in promoting community food security, particularly in low-income, urban neighborhoods.[1]

In American cities that have been especially hard hit by economic decline or that suffer from degraded environments, urban agriculture is increasingly being viewed by communities as a useful indicator of resilience.[2] Older, industrial cities—such as Cleveland, Detroit, and Buffalo—with their drastic losses of population and acres of vacant land resulting from depopulation and disinvestment, are emerging as centers for urban agriculture initiatives. In essence, they are becoming laboratories for the future role of urban food production in the postindustrial city. Yet urban agriculture is also an

Urban agriculture is increasingly seen as an indicator of community resilience.

Kimberley Hodgson

increasingly important land use in dense, built-out cities such as Seattle and New York. Problems of food access disparities, childhood obesity, and food illiteracy have prompted urban agriculture activity on a variety of traditional and nontraditional spaces on public and private property.[3]

Finally, urban agriculture is part of a larger community-based food-system continuum that spans rural, periurban (peripheral areas where urban or suburban meets rural), suburban, and urban areas. As such, it is a key component of the emerging practice area of community and regional food systems planning that appears to be garnering increased acceptance among planning practitioners, educators, and students. As described in PAS Report No. 554 (Raja et al. 2008), community food planning seeks to foster greater levels of health and nutrition, particularly in low-income communities, by creating productive "food environments" through programmatic efforts (including community gardens and urban farms, farmers markets, and direct farm-to-school meal programs), policy initiatives (food charters and food policy councils), and comprehensive plans and zoning measures that accommodate community food activities.

A community-based food-systems approach has the potential to simultaneously address issues of food security, public health, social justice, and ecological health in local communities and regions, as well as the economic vitality of agriculture and rural communities. Such an approach emphasizes, strengthens, and makes visible the relationships among producers, processors, distributors, and consumers of food at the local and regional levels (Raja et al. 2008), while aiming to be:

Place-based, promoting networks of stakeholders, linking urban and rural issues, engaging residents, and creating senses of place;

Ecologically sound, using environmentally sustainable methods for producing, processing, distributing, transporting, and disposing of food and agricultural by-products;

Economically productive, bolstering development capacity and providing job opportunities for farmers and community residents;

Socially cohesive, facilitating trust, sharing, and community building across a diverse range of cultures and addressing the concerns and needs of marginalized groups, including minority and immigrant farmers and farm laborers, financially struggling small farmers, and underserved inner-city and rural residents; and

Food secure and literate, providing equitable physical and economic access to safe, nutritious, culturally appropriate, and sustainably grown food at all times across communities and fostering an understanding and appreciation of food, from production to disposal.

While programs, projects, and entrepreneurial activity are important components of a healthy, sustainable food system, their replication and effectiveness are often hindered by the absence of public policies that provide governmental, legal, and institutional support for community-based food systems (Raja et al. 2008). Historically, planners and local governments have had limited interests in food-systems issues and food policy (Pothukuchi and Kaufman 1999, 2000; Caton Campbell 2004). However, a number of formal and informal coalitions of food-system stakeholders, including local and regional governments and planners, are developing and implementing successful plans and policies to address issues—from food production to waste disposal—in hopes of creating healthier, more sustainable food systems, communities, and people.

FRAMEWORK

This PAS Report is the latest in a series of APA education, outreach, research, and policy actions and publications related to community and regional food systems planning. In his opening keynote address at the 2003 National Planning Conference, Jerry Kaufman, FAICP, asked, Why are planners not engaged in the food system, since they are actively engaged in air, water, and shelter issues, all of which are basic necessities vital to not just the quality of life but life itself?

Planners' interests and engagements in food-systems issues began to grow not long thereafter. In 2004, special issues of the *Journal of Planning Education and Research* and *Progressive Planning* emphasized the breadth and depth of linkages between the food system and other areas of planning practice. APA's National Planning Conference also included special tracks on food planning in 2005 (San Francisco) and 2006 (San Antonio). While a few academic planning departments had made community food-systems planning part of their course offerings as early as 1997, planning programs at other schools—including the University of California at Los Angeles, the University of Wisconsin–Madison, Wayne State University, and the University of Virginia—followed suit as student interest burgeoned.

In 2005, Kaufman—along with Deanna Glosser, president and CEO of Environmental Planning Solutions and former APA Divisions Council vice chair, and Kami Pothukuchi, associate professor of urban planning and director of SEED Wayne (Sustainable Food Systems Education and Engagement in Detroit and Wayne State University)—initiated and launched the Food Interest Group (FIG), a coalition of APA members interested in or actively engaged in food-system planning at the local, regional, state, or national levels. In 2006, FIG prepared and presented a white paper on food planning to the Delegates Assembly at the National Planning Conference. Approved subsequently by the APA Legislative and Policy Committee, the white paper became the impetus for the preparation of the *Policy Guide on Community and Regional Food Planning* (APA 2007). APA has subsequently authored and published several reports and resources, which are included in the references at the end of this report.

In 2008, APA launched the National Centers for Planning, which are dedicated to helping planners create communities of lasting value: safe, healthy, and sustainable places that respect the values of their citizens. One of these, the Planning and Community Health Research Center (PCHRC), focuses on integrating community health issues into local and regional planning practices by advancing a program of policy, relevant research, and education. The PCHRC provides practicing planners and allied professionals with guidance on how to improve community and regional food systems. (See www.planning.org/nationalcenters/health/food.htm for further information.)

The events, publications, and activities outlined above are the foundation of this report. In addition, the authors and APA researchers developed case study research and conducted in-depth interviews with planners, local government officials, and urban agriculture practitioners in 11 cities across the United States and Canada. This research was designed to identify the opportunities and challenges faced by cities and counties of varying sizes, economies, and locations in supporting and expanding urban agriculture, illustrating the range of municipal efforts and variety of policies and programs both emerging and in place. The case studies also reveal differences among cities in their approaches and emphases as they respond to the needs of the urban agriculture community. The cities and regions studied were Chicago; Cleveland; Detroit; Kansas City, Kansas and Missouri; Milwaukee; Minneapolis; New Orleans; Philadelphia; Seattle and King County, Washington; Toronto, Ontario; and Vancouver, British Columbia.

AUDIENCE

In most cases, practicing planners in the private or public sectors and other local and regional government staff are not currently leading the urban agriculture movement in North America. Instead, the urban agriculture movement is being led primarily by dedicated individuals and community-based nonprofit organizations—some of which were created expressly to engage in urban agriculture, others of which added it to their menu of activities. This report is intended to encourage planners to expand their involvement in and support of urban agriculture–related policies, programs, and projects and to integrate urban agriculture into food-system planning processes.

Most planners already possess sets of skills that are relevant and applicable to the urban agriculture movement. Even without knowledge of or experience in urban agriculture, planners can apply their abilities to envision alternative urban futures, their professional knowledge of urban systems, their grasps of land-use change and regulation mechanisms (such as comprehensive plans and zoning ordinances), their abilities to facilitate collaboration within government and with nongovernmental organizations and other professionals, and their expertise in community engagement and consensus building.

This report provides a conceptual and practical guide for planners working in the public sector. Private and nonprofit-sector planners—as well as staff of other local and regional government agencies, including but not limited to public health, environment, economic development, and community development—may also find this report relevant to their work. In addition, this report should be of use in the growing number of university-level courses in food-systems planning.

The urban agriculture movement is being led primarily by dedicated individuals and community-based nonprofit organizations.

Adam Golfer

Finally, through its collection of case studies, this report serves as a snapshot of the state of urban agriculture practice in the United States and Canada. As such, it should be of interest to readers beyond the planning community. Chief among these are urban agriculture practitioners, who may interact with a variety of grassroots community food-system stakeholders outside of traditional local planning frameworks. They have long recognized the importance of planners in facilitating access to public land or other underutilized space and of policy makers who can influence the regulatory contexts in which urban agriculture operates. Thus, this report seeks to inform practitioners of public policies and planning approaches that might be applied to their communities and to help them gain clearer senses of what they can ask of their planning departments at the neighborhood, municipal, county, and regional levels.

ENDNOTES

1. Food security is defined as "a condition in which all community residents obtain a safe, culturally acceptable, nutritionally adequate diet through a sustainable food system that maximizes community self-reliance and social justice" (Community Food Security Coalition 2010).

2. Resilience describes the capacity of a city or town to thrive in the face of social, economic, or environmental challenges. A resilient city reduces its dependence on natural resources (land, water, materials, and energy) while simultaneously improving its quality of life (ecological environment, public health, housing, employment, and community) so that it can better fit within the capacities of local, regional, and global ecosystems.

3. Food literacy is the understanding of how food is produced, transformed, distributed, marketed, consumed, and disposed of.

CHAPTER 2

What Is Urban Agriculture?

 Gaining a comprehensive understanding of how planning and local government policy can support and encourage urban agriculture requires a thorough grasp of urban agriculture: its history, evolution, and current definition, as well as its various dimensions, types, benefits, risks, and prerequisites.

HISTORY OF URBAN AGRICULTURE

Agriculture has been a part of North American cities for centuries, though planning has alternately worked to promote and prevent agricultural activities in urban areas. In the colonial era, agriculture was central to urban economic growth and was promoted in early plans. Although cities industrialized in the 19th century, farms and food processing remained part of the urban landscape, and even in the 20th century urban gardening programs promoted food production as a use of vacant land and a strategy for coping with the economic challenges of war, depression, and inner-city decline. At the same time, professional planners seeking to regulate land use and improve public health increasingly defined farming as a rural activity. With attempts over the last few decades to reintroduce or scale up urban farming, planners are being challenged to define the appropriate place of agriculture in cities.

Agriculture has been a part of North American cities for centuries, though planning has alternately worked to promote and prevent agricultural activities in urban areas.

Food production was the basis of most colonial settlements' household and regional economies. The planners of early North American towns gave agriculture a central place. Boston and other New England towns reserved a "common" for farm animals to graze on. William Penn envisioned Philadelphia as a "green countrie town" with acre and half-acre lots that "hath room for House, Garden and small Orchard, to the great Content and Satisfaction of all here concerned" (see Myers 1912, 283). Outside the city, Penn and his surveyors planned a belt of large agricultural estates, beyond which early settlers established farming villages and mills to process grain and other products.

As cities industrialized in the 19th century and large-scale farming of grain and meat came to dominate the North American interior, the metropolitan geography of agriculture shifted. In the hinterlands of major cities, farmers unable to compete with bulk crops such as corn and wheat transitioned to dairy, vegetable market gardening, orchards, and other higher-value, perishable crops for urban consumers. Economic development institutions such as the Philadelphia Society for Promoting Agriculture encouraged this shift (Baatz 1985). At the same time, the expansion of public markets reduced the need for city dwellers to grow their own food. By the late 19th century, though some farms still remained in cities, urban agriculture was becoming less a necessity and more a form of private recreation as well as a resource for charity. The great urban parks developed by the early professional planners and landscape architects included distinct agricultural features, sometimes in the form of pasture for grazing animals or dairies that supplied milk to young children and mothers (Figure 2.1; Rosenzweig and Blackmar 1992).

Figure 2.1. Central Park's Dairy Visitor Center was originally used as a dispensary to provide milk to families.

The depression that followed the financial panic of 1893 inspired the first large-scale urban agriculture programs intended to address poverty and economic need. In 1894, high unemployment in Detroit led Mayor Hazen S. Pingree to initiate, despite widespread skepticism, a garden program on vacant land being held for speculative purposes. Pingree's "potato patches" were envisioned as supplements to existing charity efforts. Within two years, almost half of Detroit's families on public relief were growing food on lots of various sizes, most of them at the edge of town. The Detroit experiment was quickly replicated elsewhere. Philadelphians established the Vacant Lot Cultivation Association, promoting market gardening as well as production for household consumption. An 1898 report by the New York Association for Improving the Condition of the Poor reported similar programs in 19 cities (Lawson 2005). Similarly, settlement houses in inner-city neighborhoods made gardening and food part of social reform and community development beginning in the late 19th century. Workers at these neighborhood centers planted vegetable gardens and ran cooking and food-processing programs that provided relief for poor families while also orienting new immigrants to American "food ways" (Vitiello, forthcoming). Settlement houses also helped mobilize city residents to scale up gardening during the 20th century's world wars and the Great Depression.

While such small-scale urban agriculture efforts grew, professional planners at the beginning of the 20th century saw more intensive agricultural uses—such as animal production and meat processing—as threats to public health and safety, and they used the new tool of zoning to move such facilities out of central cities. At the same time, planners were concerned with ensuring safe and adequate food supplies, producing reports on regional production, transportation, and wholesale markets in what became known as metropolitan "foodsheds" (Donofrio 2008).[1] Their concerns are echoed today in planning and policy reports on food safety, food security, and farmland preservation.

The early 20th century also saw the U.S. government become involved in the urban agriculture movement. In response to food shortages during World Wars I and II and the need to boost public morale, it encouraged rural and urban Americans to plant victory gardens, also known as "war gardens" or "food gardens for defense." (See Figure 2.2.) Similar programs addressed the crisis of the Great Depression in the 1930s. Victory and Depression gardens were the largest-scale urban agriculture initiatives in the United States to date. In 1943, more than 20 million gardens sprouted on private and public land—in front lawns, backyards, and public parks, and on empty lots and rooftops—producing an estimated 9 to 10 million tons of fruits and vegetables, or about 41 percent of all vegetable produced that year (Reinhardt n.d.). A handful of war gardens survive today, including Boston's Fenway Victory Gardens. Once used by more than 2,500 families, it is now a high-profile, 500-plot community garden on city parkland (Kaufman and Bailkey 2000).

The economic boom and accelerated city and suburban growth following World War II pushed agriculture even farther from cities. By the mid-20th century, many cities' zoning codes no longer included farming as a recognized land use; residential development had claimed most former farmland inside cities, and modernist planners did not see agriculture as part of city life. The industrialization of farming with its chemical fertilizers, pesticides, and preservatives helped make food supply chains international, further diminishing the role of local agriculture in feeding urban residents. Yet at the same time, the decline of many inner cities inspired new generations of urban agriculture. Some social programs confronting disinvestment, racial

Figure 2.2. *During World War II, the federal government exhorted citizens to participate in the victory gardens program.*

change, and other aspects of the urban crisis of the postwar decades sometimes promoted gardening, such as Philadelphia's Neighborhood Gardens Association, established in 1953. The current grassroots-driven urban agriculture movement took shape in the 1970s, as people planted community gardens in major metropolitan areas across the country.

Community gardens were responses to deindustrialization, depopulation, increases in acreage of vacant land, and the failures of urban renewal but also to immigration. In most northern cities, the largest number of gardeners were African-Americans from the South; in some cities, Puerto Ricans, as well as Southeast Asians resettled following the Vietnam War, brought agricultural knowledge and skills to inner-city neighborhoods. Older immigrants also planted garden plots, as did more affluent whites in an early era of central-city gentrification (Vitiello and Nairn 2009). In Boston, by the middle of the 1970s, gardeners—some of whom had gained political experience by opposing urban renewal projects—initiated a series of unauthorized appropriations of vacant parcels. The founders of Boston Urban Gardeners (BUG) in 1977 realized that the initiation, expense, and future of garden sites would be limited without processes that engaged both neighborhood and citywide politics (Warner 1987).

Government and nonprofit programs institutionalized the community gardening movement to varied extents. Between 1977 and 1996, the U.S. Department of Agriculture started an Urban Gardens Program in which agricultural extension agents across the country supported city residents in developing and sustaining gardens, providing seeds and technical advice. In many cities, local philanthropists funded new programs that provided material support for gardens, including fences, compost, and wood for raised beds. By the mid-1990s, New York City and Philadelphia each claimed more than 1,000 gardens providing food and spaces for ornamental plantings. The members of the San Francisco League of Urban Gardeners (SLUG) adapted the BUG model, acting as garden creators, sources of technical gardening assistance, and political advocates. At the national level, the American Community Gardening Association, founded in 1979, promoted information sharing and garden advocacy actions. In cities such as Chicago, New York, and Boston, planners and redevelopment agency staff supported gardeners in accessing land and in some cases preserving its use in agriculture.

As cities recovered and real-estate development pressures began to grow, community gardening in many cases became an activist cause. As the market heated up in New York City in the 1990s, public and private interests—and planners—promoting redevelopment saw established gardens as interim land uses. The city subsequently lost hundreds of community gardens to development. Philanthropic support for community gardening also waned, as foundations erroneously concluded that they were doing little more than funding a middle-class hobby. In Philadelphia, as foundation funding for garden support programs dried up and older gardeners died, the number of gardens in the city plummeted. Between 1996 and 2008, the number of food-producing community gardens in the city declined from 501 to 226 (Vitiello and Nairn 2009).

In some cities, this instability led public and private institutions to focus on land preservation. Land trusts such as the Southside Community Land Trust (founded 1981) in Providence, Rhode Island, identified and protected land for permanent or long-term use as community gardens and later urban farms. The Neighborhood Gardens Association in Philadelphia, recognizing community gardens' positive impacts on neighborhood revitalization and public space, acquired properties mainly in gentrifying areas surrounding the downtown, where residents lacked large yards. Since 1974, the City of Seattle has overseen the P-Patch system of community gardens, which today boasts more than 70 garden sites and 2,000 plotholders. (See Figure 2.3.) In Chicago, the city government invested in Neighbor Space (1996), a land trust that has preserved more than 60 community gardens. A public agreement among New York City's department of parks, department of housing preservation and development, and community gardeners stabilized many gardens following the increase in development of the 1990s. However, most city governments have taken a less systematic approach to land-use planning and preservation for urban agriculture, issuing temporary use permits for gardens and farms on publicly held vacant land and viewing agriculture as an interim use.

Figure 2.3. The City of Seattle oversees more than 70 community gardens in its P-Patch program.

Kimberley Hodgson

Today, urban agriculture is guided largely by local, city-based organizations operating as tax-exempt nonprofits. They include citywide garden-support programs that provide tools, seedlings, and other materials to backyard and community gardeners; youth programs that teach about food and nutrition; business-incubator farms and training programs that enable refugees and aspiring farmers to gain farming and marketing skills; and networks of small commercial growers. This new generation of urban farms challenges planners and city governments to determine appropriate land-use planning and regulations that can support urban agriculture in the 21st century.

DEFINING URBAN AGRICULTURE

Urban agriculture encompasses far more than private and community gardens. It is typically defined as the production of fruits and vegetables, raising of animals, and cultivation of fish for local sale and consumption. A more holistic systems definition acknowledges the connection between urban agriculture and the larger food system, as well as its influence and dependence on a variety of economic, environmental, and social resources.

How urban agriculture is defined varies broadly by region and country, as well as by field of study. In the past five years, however, these definitions have grown to encompass much more than simply the production of food within urban areas.

In 2007, the Community Food Security Coalition's Urban Agriculture Committee established a comprehensive definition of urban agriculture to address its multiple dimensions and forms of practice:

> Urban and peri-urban agriculture (UPA) refers to the production, distribution and marketing of food and other products within the cores of metropolitan areas (comprising community and school gardens; backyard and rooftop horticulture; and innovative food-production methods that maximize production in a small area), and at their edges (including farms supplying urban farmers markets, community supported agriculture, and family farms located in metropolitan greenbelts). Looked at broadly, UPA is a complex activity, addressing issues central to community food security, neighborhood development, environmental sustainability, land use planning, agricultural and food systems, farmland preservation, and other concerns. (Community Food Security Coalition 2007)

Urban agriculture is embedded in communities . . . yet it is part of the larger food-system continuum.

As this definition indicates, urban agriculture is embedded in communities. Yet it is part of the larger food-system continuum, including not only the production of food within urban, suburban, and rural built environments but also its related physical and organizational infrastructure and associated policies and programs. (See Figure 2.4.)

Figure 2.4

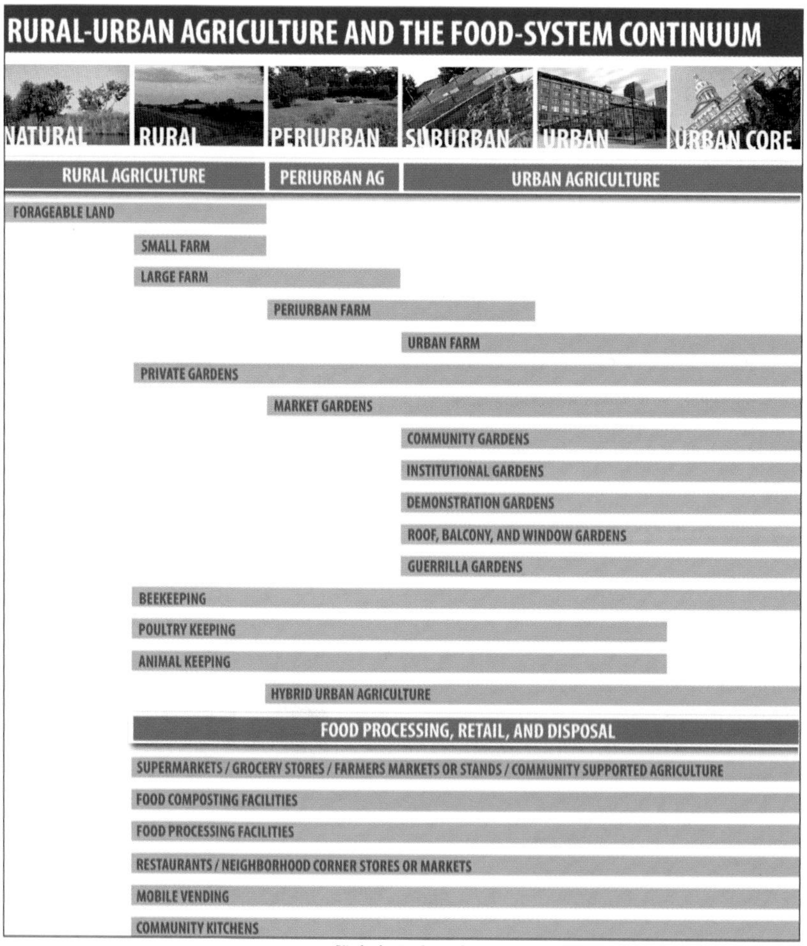

RURAL-URBAN AGRICULTURE AND THE FOOD-SYSTEM CONTINUUM

NATURAL — RURAL — PERIURBAN — SUBURBAN — URBAN — URBAN CORE

Kimberley Hodgson, from a concept by Andrés Duany; design by John Reinhardt

Dimensions of Urban Agriculture

Besides community and private vegetable gardens, other important but less-common types of urban agriculture include institutional and demonstration gardens; edible landscaping; hobby and commercial bee-, poultry, and animal keeping; urban and periurban farms; and hybrid forms that integrate gardening and farming activities for personal consumption, educational purposes, donation, or sale. There is considerable variation in the purpose, location, size and scale, production techniques, and end products of these and other types of urban agriculture.

Purpose. Urban agriculture can produce plants or animals for personal consumption or use, educational or demonstration purposes, neighborhood revitalization or economic development, healing or therapeutic purposes, sale or donation, or a combination of the above.

Location. Urban agriculture activities (including the production, processing, and sale of plants, animals, and ornamentals) can be located within an urban, suburban, or periurban area, on underutilized private or public land, spaces, or on building sites in developed residential, commercial, or industrial areas.

Size and Scale. Urban agriculture can occur almost anywhere: on large, contiguous parcels of land; on small, noncontiguous parcels of land; or in other spaces such as rooftops, balconies, porches, utility rights-of-way, fences, walls, or basements. (See Figure 2.5.)

Production Techniques. Urban agriculture can utilize a variety of production techniques, such as in-soil or raised-bed cultivation, hoop house or greenhouse growing, hydroponics, aquaponics, permaculture, or vertical farming. (See Figure 2.6.)

End Products. Urban agriculture can include the production of plants or animals for consumption or ornamental use, as well as the production of key urban agriculture inputs, such as compost.

Typology of Urban Agriculture

Urban agriculture can take many forms but can be broadly classi-

Figure 2.5. Urban agriculture can occur in almost any space, including along fence lines.

Figure 2.6. Hydroponics is one of many production techniques that can be used by urban growers.

fied according to one of three categories: noncommercial, commercial, or hybrid.

Noncommercial types include private, community, institutional, demonstration, and guerrilla gardens; edible landscaping; and hobby bee- and chicken keeping. The Edible Schoolyard is a well-known example of an institutional garden. This one-acre organic garden is located on the property of Martin Luther King Jr. Middle School in Berkeley, California, and provides hands-on gardening, science, nutrition, and ecology education to students. First Lady Michelle Obama's White House Garden, Baltimore mayor Sheila Dixon's City Hall vegetable garden, and the San Francisco City Hall Victory Garden are only a few examples of the demonstration gardens appearing in cities across the country to show that urban agriculture can contribute to health, social, economic, and environmental goals.

Commercial types include market gardens; urban and periurban farms; beekeeping operations; aquaponic and hydroponic systems; and the equipment, materials, and structures required to process, distribute, and sell food (plant, animal, or fish) products. Potomac Vegetable Farms (PVF), founded in 1960, is a commercial agriculture operation, located on a 10-acre urban farm in Vienna, Virginia, and a 180-acre periurban farm in Purcellville, Virginia (www.potomacvegetablefarms. com). PVF's primary source of revenue is a 450-member community-supported agriculture program, or CSA, with weekly shares of varying sizes. CSA members include individuals and businesses in Alexandria and Arlington, Virginia, as well as Washington, D.C. Other sources of revenue include sales from several farmers markets and an on-site farmstand at the Vienna location. PVF employs three full-time staff and several part-time and seasonal employees.

Hybrid types, often referred to as social enterprises, include any combination of food production, processing, distribution, marketing, or educational activities and are typically operated by a nonprofit organization for social, economic, or environmental purposes. Lynchburg Grows (Lynchburg, Virginia), Kansas City Community Farm (Kansas City, Kansas), Earthworks Urban Farm (Detroit), Green Youth Farm (Chicago), Red Hook Community Urban Farm (New York), Growing Power (Milwaukee), and Hollygrove Market and Farm (New Orleans) are all examples of this emerging form of urban agriculture. In addition to producing food for sale at a number of retail destinations—including on-site farm stands, community farmers markets, CSAs, and locally owned and operated food retail businesses—they offer a range of community and educational programs for children, youth, adults, and specific populations, such as homeless people, pregnant teens, and formerly incarcerated youth or adults.

Table 2.1 and its visualization define and provide examples of these various categories and types of urban agriculture and their related dimensions. Seen together, they form a spectrum from a collection of pots on a balcony to a multi-acre farming operation no different in appearance and intent from traditional agriculture.

The viability of urban agriculture, however, depends heavily on specific infrastructure: uncontaminated growing media (soil, compost or water), accessory structures and materials, and, for commercial urban agriculture, processing facilities, distribution channels and equipment, and direct-sale retail destinations. Table 2.2 classifies and describes this urban agriculture infrastructure in more detail.

CATEGORY	TYPE	DESCRIPTION
NONCOMMERCIAL	Private Garden	Private food-producing gardens located in the front or back yard, rooftop, courtyard, balcony, fence, wall, window sill, or basement of a private single-family or multifamily residence, attended to by an individual or gardening business. End products are typically used for personal consumption. Examples: National Gardening Association (www.garden.org), American Horticultural Society (www.ahs.org), Organic Gardening (www.organicgardening.com)
	Community Garden	Small- to medium-scale production of food-producing and ornamental plants, on contiguous or discontinuous plots of land, located on public or private property in residential areas, gardened and managed collectively by a group. Gardening activities and end products are typically used for consumption or education; however, they may also be sold on- or off-site, depending on local government regulations and the goals of the garden as a collective effort. Examples: American Community Gardening Association's community garden database (http://acga .localharvest.org), P-Patch Community Gardens (Seattle; www.seattle.gov/neighborhoods/ ppatch), Neighborhood Gardens Association (Philadelphia; www.ngalandtrust.org)
	Institutional Garden	Small to large food-producing gardens or orchards located on private or public institutional property (school, hospital, faith-based organization, workplace) in a residential, commercial, or mixed-use area, gardened by an organization or business. The process of gardening is typically used for educational, therapeutic, and community service purposes—including but not limited to nutrition education, environmental stewardship, and community ministry. The end products are typically used for donation or consumption. Depending on local government regulations, they may also be sold on- or off-site at a stand, market, or store to financially support the garden's specific activities. Examples: Edible Schoolyard garden (Berkeley, Calif.; www.edible schoolyard.org), Google Corporation organic garden (Mountain View, Calif.; www.google .com/corporate/green/employee-benefits.html), Harvard Pilgrim Health Care employee garden (Wellesley, Mass.), Legacy Good Samaritan Hospital garden (Portland, Ore.); Sophia Louise Durbridge-Wege Living Garden at the Family Life Centre (Grand Rapids, Mich.)
	Demonstration Garden	Small food-producing garden located on private property (school, hospital, faith-based organization, workplace) or public property (park, school, and other civic space) in a residential, commercial, or mixed use area for public demonstration purposes only, gardened by a local government agency, community organization, or business. End products are typically donated to local organizations and food banks. Examples: Baltimore City Hall vegetable garden, San Francisco City Hall Victory Garden, Not a Cornfield (Los Angeles; http://notacornfield.com), Public Farm One (New York; www.publicfarm1.org)
	Edible Landscape	The use of food-producing plants in the design of private and public outdoor spaces in residential, commercial, and mixed use developments, attended to by an individual or business. End products are typically used for consumption. Examples: Edible Estates (www.fritzhaeg.com/ garden/initiatives/edibleestates/main.html), South East False Creek Mixed Used Development (Vancouver, B.C.; http://vancouver.ca/commsvcs/southeast/docments/pdf/designingUA.pdf)
	Guerrilla Garden	Unauthorized appropriation and cultivation of food-producing or ornamental plants on untended, abandoned, or vacant private or public land by an individual or group. End products are typically used for neighborhood revitalization purposes. Examples: Los Angeles Guerrilla Gardening (www.laguerrillagardening.org), SoCal Guerrilla Gardening (http:// socalguerrillagardening.org), Edmonton Guerrilla Gardening (http://edmontongg.blogspot. com), South Phila Guerrilla Gardening (http://guerrillaphilly.wordpress.com), Green Guerillas (New York; www.greenguerillas.org)
	Hobby Beekeeping	Small-scale keeping of honeybees for personal use. Beehives can be colocated with gardens or nongarden uses (such as parks), on underutilized spaces (including rooftops) in residential, mixed use, or other public land areas. End products are typically used for personal consumption, education, or donation. Examples: City Hall Bees (Vancouver, B.C.; http:// vancouver.ca/commsvcs/socialplanning/initiatives/foodpolicy/projects/beekeeping.htm), New York City Beekeepers Association (www.nyc-bees.org)
	Hobby Chicken Keeping	Small-scale keeping of chickens for personal use in residential areas, or for commercial use in residential, mixed use, or other public land areas. Poultry keeping can be colocated with other agriculture and nonagriculture uses. End products are typically used for personal consumption, education, or sale. Examples: Backyard Chickens (Vancouver, B.C.; http:// vancouver.ca/commsvcs/socialplanning/initiatives/foodpolicy/projects/chickens.htm), A2 City Chickens (Ann Arbor, Mich.; www.a2citychickens.com), Chicken Keepers (Cleveland; www.localfoodcleveland.org/group/chickenkeepers)

Table 2.1. *Typology of urban agriculture*

(continued)

(continued from page 17)

CATEGORY	TYPE	DESCRIPTION
C O M M E R C I A L	**Market Farm**	Small- to medium-scale production of food-producing or ornamental plants, bees, fish, poultry, or small farm animals located on public or private property, and designed and managed for commercial purposes using a variety of growing techniques including in-soil, container, hydroponic, and aquaponic growing systems. End products are typically sold on- or off-site at a stand, market, or store. Examples: Urban Growth Farm (Cleveland; www.urbangrowthfarms .com), Fresh Roots Farm (Atlanta; www.freshrootsfarm.com)
	Urban Farm	Typically larger than market gardens and include larger-scale production of food-producing or ornamental plants, bees, fish, poultry, or small to medium-sized farm animals for commercial purposes using a variety of horizontal and vertical growing techniques including in-soil, container, hydroponic, and aquaponic growing systems. End products are typically sold on- or off-site at a stand, market, or store. If large enough, urban farms may adopt the community-supported agriculture (CSA) distribution model, through which consumers of the farm's produce over the growing season also share in its risks. Examples: Greensgrow Farm (Philadelphia; www .greensgrow.org), Red Planet Vegetables (Providence, R.I.; http://redplanetvegetables.wordpress .com), Springdale Farm (Austin, Tex.; http://springdalefarmaustin.com), Brooklyn Grange Farm (Brooklyn, N.Y.; http://brooklyngrangefarm.com)
	Periurban Farm	Practiced outside or on the fringes of metropolitan areas, often on agricultural land facing some threat of future development. Includes larger-scale production of food-producing or ornamental plants, bees, fish, poultry, or small to large farm animals for commercial purposes using a variety of growing techniques including in-soil, container, hydroponic, and aquaponic growing systems. Such farms are managed as agricultural businesses and may employ organic techniques or the CSA model. In most cases, the farm's production is marketed and distributed in the nearby metropolitan area. Examples: Potomac Vegetable Farms (Vienna, Va.; www .potomacvegetablefarms.com), Full Circle Farm (King County, Wash; www.fullcirclefarm.com)
	Beekeeping	Medium- to large-scale keeping of honeybees for commerical use. Beehives can be colocated with other urban agriculture uses (such as market gardens or urban farms) or other nonagriculture uses (such as parks or rain gardens), on underutilized spaces (including rooftops), in residential, commercial, mixed use, or industrial areas. End products are typically used for sale. Examples: Backyard Bees (Southern Calif.; http://backyardbees.net), Burgh Bees (Pittsburgh; www .burghbees.com), Earthworks Urban Farm (Detroit; www.cskdetroit.org/EWG/apiary.cfm).
H Y B R I D	**Hybrid Urban Agriculture**	Any combination of gardens and farms that produce food-producing or ornamental plants, bees, fish, poultry, or small to medium-sized farm animals for personal consumption, education, donation, and sale. Examples: 21 Acres (King County, Wash.; http://21acres .org), Hollygrove Market and Farm (New Orleans; www.hollygrovemarket.com), Growing Power (Milwaukee, Wis.; www.growingpower.org), Lynchburg Grows (Lynchburg, Va.; www .lynchburggrows.org), GROWHAUS (Denver; www.thegrowhaus.com)

Note:

Small = 0 to ½ acre or 1 beehive, 1–4 poultry, or 1 animal.

Medium = 1 to 2 acres or 2–4 beehives, 5–10 poultry, or 2–4 animals depending on poultry or animal size and available space.

Large = 5–10 beehives, 11 or more poultry, or 5–10 animals depending on poultry or animal size and available space.

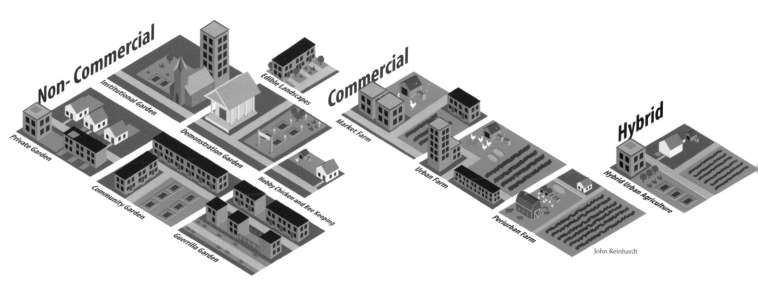

John Reinhardt

ELEMENT	DESCRIPTION
Accessory Structures and Materials	
Growing	Raised beds, containers, and similar contained growing systems; planting-preparation houses or similar structures; greenhouses, hoop houses, coldframes, and similar structures used to extend the growing season; or hydroponic equipment, supplies, and structures
Irrigation	Water hoses, rain barrels, and other equipment used to irrigate the garden or farm
Compost	Bins, worms, screens, inputs (household, restaurant, or other food-service food waste; yard wastes; and poultry or animal manure), and other materials
Bees, poultry, and animals	Beehives, coops, cages, stables, barns, fences, or other enclosures
Fish	Aquaponic equipment, supplies, and structures
Storage	Tool sheds, dry or cold storage rooms, and other similar structures
On-site sales	Farm stand, retail store, or similar structure
Other	Benches, shade pavillions, restroom facilities, office space, picnic tables, children's play areas, and other structures and spaces.
Processing Facilities	
On-site facility	Partially or fully equipped kitchen for food preparation, preservation, or packaging, located on-site for personal consumption or commercial purposes. Commercial facilites are state inspected and licensed to allow the preparation and preservation of food for sale to a variety of retail destinations.
Community kitchen	Shared-use facility with a partially to fully equipped kitchen used for food preparation, preservation, and packaging. Noncommercial facilities are used for personal consumption purposes only and can be located anywhere from church basements to community centers to freestanding structures. Commercial facilities are state inspected and licensed to allow the preparation and preservation of food for sale to a variety of retail destinations.
Community processing center	Small-scale state-inspected and licensed facility containing a variety of equipment, which can be rented by urban growers to add value to raw food products through processing, packaging, and subsequent delivery to retail destinations
Distribution	
Food hub	Centrally located facility with a physical drop-off point for multiple food producers (gardeners, farmers) and a pick-up point for food buyers (restaurants, stores, institutions, cooperatives, caterers, etc.) wanting to buy locally grown or raised food products; or an online, virtual meeting place to connect food producers and sellers with food buyers
Retail Destinations	
Farm stand	Small retail venue, typically featuring one urban farmer, located on-site at a market garden, urban farm, or periurban farm, to sell agricultural products directly to consumers
Farmers market	Retail venue featuring multiple urban, periurban, or rural farmers operating within a certain geographic area to sell agricultural products directly to consumers
Community-supported agriculture	Direct grower-to-consumer sale and distribution model that emphasizes shared investment, responsibility, and risk. A grower sells a share of farm output to individuals and families at the beginning of the growing season and supplies seasonal produce and other agriculture products weekly or biweekly throughout the growing season. Agricultural products are typically distributed directly from the farm to an individual shareholder's home, place of work, or designated pick-up site.
Farm-to-institution	Direct sale of locally produced food products to schools, universities and colleges, hospitals and long-term care facilities, prisons and correctional facilities, and other institutional facilities
Food cooperative	Member-owned, member-controlled food business made up of food producers and consumers. Facilitates the direct sale and purchase of agricultural products between members at a designated store; members may be required to pay an equity investment to join the co-op or work in the store, and in return receive special benefits, such as reduced rates
Other	Restaurants, catering businesses, corner stores, bodegas, mobile food carts, and small and larger grocery stores

Table 2.2. Urban agriculture infrastructure

Urban agriculture provides opportunities for community involvement and social interaction among ethnically and age-diverse communities.

Kimberley Hodgson

BENEFITS OF URBAN AGRICULTURE

Urban agriculture helps meet local food needs while promoting environmental sustainability, health, nutrition, and social interaction; creating opportunities for locally controlled food enterprises and economic development; and enhancing community engagement and empowerment. The sections below detail its many benefits.

Health Benefits

According to recent research, urban agriculture can increase access to fruits and vegetables, especially in low-income areas that have limited access to affordable, healthful foods. Urban agriculture also provides opportunities for public health programming to improve nutrition knowledge, attitudes, and dietary intake (Bellows, Brown, and Smit 2004; McCormack et al. 2010). DeLaney Community Farm, a project of Denver Urban Gardens located in Aurora, Colorado (http://dug.org/delaney), was established to improve access to fresh fruits and vegetables and provide opportunities for nutrition education for low-income and other area residents. Courses focus on the entire experience of food production and consumption: growing, preparing, cooking, tasting, and sharing. Alemany Farm, a 4.5-acre hybrid urban farm in southeastern San Francisco (www.alemanyfarm.org), works to increase food security and provide environmental education opportunities for residents through workshops, educational courses, a free neighborhood produce-delivery program, and field trips. The Community Action Coalition (CAC) of South Central Wisconsin, a nonprofit antipoverty organization in Madison (www.cacscw .org/gardens), supports low-income families in growing food in more than 50 community gardens throughout the city. CAC reports that their efforts have provided nearly 2,000 low-income households with improved access to fresh fruits and vegetables, reducing their expenditures on food.

Urban agriculture can also offer therapeutic benefits. Renewal Farm, a periurban farm 50 miles north of Manhattan in Garrison, New York, helps rehabilitate recovering drug addicts and alcoholics from New York City. Located adjacent to a rehabilitation center, Renewal Farm provides routine, structure, and opportunities for reflection to more than 24 men at a time for a period of six to nine months while they produce a variety of vegetables for area restaurants (Buckley 2009). Growing Home, a hybrid farm in Chicago (www.growinghomeinc.org), provides job training and employment for homeless and low-income individuals.

Social Benefits

Community and school gardens, hybrid urban agriculture, and direct marketing strategies (such as community-supported agriculture, farm-to-school programs, and farmers markets) provide opportunities for community involvement, social interaction among ethnically and age-diverse communities, and health and environmental-stewardship education. Direct marketing strategies in particular can foster connections between farmers and consumers and can contribute to community economic security (National Research Council 2010, 7). Urban agriculture can foster community building, mutual trust, sharing, feelings of safety and comfort, and friendships that translate to a collective investment in the common good of a neighborhood. It can also serve as an alternative vacant-property reuse strategy to decrease or prevent crime, trash accumulation, illegal dumping, littering, and fires, and as a catalyst for additional community development activities and positive place-based programs (Lyson 2005; Teig et al. 2009; Schukoske 1999; Bellows, Brown, and Smit 2004; Mallach 2006; Kaufman and Bailkey 2000; Veenhuizen 2006).

The Urban Oak Organic Farm in New Britain, Connecticut, provides good examples of many of these benefits. Located on a redeveloped brownfields site in a former manufacturing district, the farm not only contributed to the revitalization of the surrounding neighborhood but continues to provide "education for residents and school groups in organic gardening methods, sustainable agriculture, non-toxic farming techniques, composting, and other environmentally-friendly farming techniques." The farm also operates a farmers market, which further contributes to an improved community life in a once underserved, blighted neighborhood (Hersh et al. 2010).

Economic Benefits

Urban agriculture presents many economic opportunities. It can contribute to decreasing public land-maintenance costs, increasing local employment opportunities and income generation, capitalizing on underused resources (e.g., rooftops, roadsides, utility rights-of-way, vacant property), increasing property values, and producing multiplier effects through the attraction of new food-related businesses, including processing facilities, restaurants, community kitchens, farmers markets, transportation, and distribution equipment (Veenhuizen 2006; Mallach 2006; Kaufman and Bailkey 2000). Subsistence production can also reduce food expenditures and make household income available for other purposes. For example, community and squatter gardens in Philadelphia produced approximately $4.9 million in summer vegetables in 2008—an amount greater than the combined sales of all of Philadelphia's farmers markets and urban farms (Vitiello and Nairn 2009).

A recent study of New York City community gardens found that within five years of a community garden's opening, neighboring property values increased by as much as 9.4 percent and continued to increase over time. Disadvantaged neighborhoods experienced the greatest increase in property values. Furthermore, the study found that community gardens can also lead to "increases in tax revenues of about half a million dollars per garden over a 20-year period" (Voicu and Been 2008). A different study assessing the neighborhood effects of 54 community gardens in St. Louis, Missouri, found that median rent and median housing costs (mortgage payments, maintenance costs, and taxes) for owner-occupied housing, as well as home ownership rates, increased in the immediate vicinity of gardens relative to the surrounding census tracts, following a garden opening (Voicu and Been 2008, Tranel and Handlin 2006).

Environmental Benefits

Urban agriculture can contribute to environmental management and the productive reuse of contaminated land. As a result of increased plant foliage, urban agriculture can contribute to decreased stormwater runoff and air pollution, and it can increase urban biodiversity and species preservation (Kaufman and Bailkey 2000; Mallach 2006; Veenhuizen 2006). Cleveland, Ohio—site of approximately 3,300 acres of vacant land and 15,000 vacant buildings—recently completed a sustainability plan to productively reuse those properties through a variety of creative strategies, including urban agriculture. Since the plan's development, more than 30 urban-agriculture reuse projects have been implemented throughout the city.

Unfortunately, with a few exceptions, these benefits are not widely quantified or analyzed. As interest in urban agriculture continues to grow, local governments can play an important role in documenting these benefits and partnering with local colleges and universities to further research urban agriculture's impacts on communities.

RISKS OF URBAN AGRICULTURE

Urban agriculture also presents potential health and environmental risks. Important factors to consider include former uses of urban agriculture sites as well as their proximities to industry, automobile traffic, and other pollutants. (See Figure 2.7.) Soil and water may be contaminated with industrial wastes and pollutants such as heavy metals (lead, cadmium, chromium, zinc, copper, nickel, mercury, manganese, selenium, and arsenic), acids or bases, asbestos, solvents or combinations of contaminants, or pathogenic organisms. Inadequate assessment, cleanup, or containment of a site can pose serious health problems to both producers and consumers through contact with contaminated water and soil or consumption of contaminated foods (Tixier and Bon 2006).

Urban agriculture uses may also cause land-use conflicts. Inappropriately managed compost facilities, use of chemical fertilizers and pesticides, and poor cultivation, harvesting, and poultry- and animal-keeping practices can lead to noise and odor nuisances. Regulatory structures can help ensure that urban agriculture complements other land uses. Limited land tenure, access to water, and inadequate funding can also present significant challenges and risks to the success of urban agriculture. This is explored in greater detail in the following section.

Figure 2.7. Some urban garden sites, such as the League Park Market Place in Cleveland, may pose risks due to their proximity to industrial sites or automobile traffic.

Helen Liggett

PREREQUISITES FOR URBAN AGRICULTURE
The success of urban agriculture, like that of traditional rural agriculture, is dependent on a variety of factors (Tixier and Bon 2006; Veenhuizen 2006):

- Climate
- Weather
- Light
- Insects and pests
- Land or other growing space
- Secure land tenure
- Healthy, uncontaminated soil or other growing medium
- Water
- Labor
- Capital and operating funds
- Financial and technical assistance
- Agricultural skills and knowledge
- Processing and transportation infrastructure
- Distribution channels
- Consumer demand
- Viable markets

Planners and other local government staff can have considerable influence over a number of these essential resources, including access to public land and other forms of growing space, land tenure, and land-use policies; the size, scale, and siting of commercial and noncommercial urban agriculture operations; opportunities for financial and technical assistance; and the provision of educational programs to increase agricultural skills and knowledge. They can also influence the development of production, processing, distribution, and transportation infrastructure; consumer demand; and viable markets through public policies and programs.

Land and Other Growing Space

Land—a place to grow food—is a primary requirement for agriculture. Periurban agriculture in or near metropolitan areas is typically practiced on agricultural lands threatened by urban development. Urban agriculture, however, can flourish on any number of sites and spaces. As planners and practitioners consider how best to support urban agriculture, they should be cognizant of (1) the availability of growing space and land; (2) land tenure; and (3) location, siting, and land use.

Availability of Growing Space and Land. An important determinant of urban agriculture's long-term success is the availability of and access to space for food production and processing purposes (Mubvami and Mushamba 2006). In some municipalities, vacant property may be plentiful; however, it may not be immediately available, or it may be only temporarily available. In postindustrial cities, vacant property is often owned by an absentee private owner and saddled with encumbrances, such as back taxes, liens, and unpaid utility bills. These create barriers to the reclamation of land for agricultural use (Schukoske 2000).

In built-out municipalities, vacant land may be nonexistent or reserved for other uses, thus limiting the possibilities for urban farms or larger market gardens. Given the wide diversity of urban agriculture (see Table 2.1, pp. 17–18), however, it can be easily adapted to variably sized spaces in many different locations. Unlike rural agriculture, urban agriculture does not require large plots of contiguous land to be productive and successful. Land areas of 1,000 square feet and fewer have been put to productive urban agricultural use (Mougeot 1999). These include private spaces such as windowsills, containers, fences, rooftops, basements, walls, front lawns, and backyards; public land including space surrounding government buildings, parks and other open spaces, and utility and transportation rights-of-way; and underutilized institutional land on hospital grounds, school yards, university campuses, and church grounds. (See Figure 2.8.)

Figure 2.8. *In built-out municipalities, urban agriculture can be sited on public or institutional grounds, like this garden at the University of Washington.*

Kimberley Hodgson

The combination of advances in technology and creativity of agricultural methods can transform unsightly small land areas into horticultural oases (Sullivan 2006). For example, the Philadelphia Water Department invested in the development of Somerton Tanks Farm, to test the economic feasibility of converting underutilized city land to a profitable commercial urban farm. Using an intensive agricultural method developed called SPIN (Small-Plot INtensive)—which focuses on growing multiple high-value crops intensively on plots of one half-acre or less—Somerton Tanks Farm growers generated more than $68,000 in gross sales from a half-acre of land in North Philadelphia in their fourth year of operation. In a space approximately one-third the size of a football field, Somerton Tanks Farm grew more than 100 varieties of 50 different types of vegetables in 280 beds. (See www.somertontanksfarm.org.)

Though large, contiguous parcels of land can be more desirable for urban farms, they may be hard to come by or require deliberate land assembly by municipal governments. Working several noncontiguous plots of land, however, can provide more direct access to niche markets (such as the growing of herbs across the street from a restaurant) and ensure stability in the face of eviction from any particular site or crop losses due to theft or other hazards. The diversification of resources over multiple locations may offer security for an urban farmer (Mougeot 1999). The founders of SPIN—Wally Satzewich and Gail Vandersteen of Saskatoon, Saskatchewan—farm 25 noncontiguous plots of land, totaling about half an acre. They say that this method "fosters self-reliance in an urban setting ... [and] makes it more feasible to utilize organic household kitchen wastes." Their proximity to their customers—urban consumers—gives them easy access to several direct-sale markets, including individual home deliveries and two farmers markets. (See www.marketgardening.com/wallysmarketgarden.)

Land Tenure. Land tenure, or the length of time and conditions (ownership, lease, occupation, or stewardship) under which a given plot of land is available for urban agricultural use, greatly affects the level of investment made by a farmer or gardener. Outright ownership is preferred, but because land values can be prohibitively high, even in economically distressed cities, many urban farmers and community gardeners instead lease land or acquire temporary user permits from public or private organizations, such as local or state governments or land trusts. Long-term agreements (such as a 99-year lease) provide the greatest sense of security. Those made with local or state governments often include certain advantages, such as access to technical assistance and water, tools, compost, mulch, and other materials.

When long-term leases are not an option, urban agriculture practitioners are often offered short-term agreements. However, these can be revoked at any time at a landowner's discretion, with as little as 30 days' notice (Schukoske 2000; Brown and Carter 2003). In other cases, agreements may not be established at all, especially when landowners are absent or indifferent. The most notable recent example of this is the former 14-acre South Central Farm in Los Angeles. In 2006, after over a decade of gardening, approximately 350 primarily Latino farming families were evicted by the owner of the site in preparation for a new industrial development, consistent with the industrial land-use designation of the site.

Urban farmers and community gardeners who use untended property without a lease or other agreement are known as squatters. These growers face additional barriers such as trespassing violations, lack of water access, and no liability insurance (Schukoske 2000). "Guerrilla gardeners" also operate without permission, but their mission concerns larger social benefits to the community: they seek to beautify and green neglected private and public plots of land (Mooallem 2008). Guerrilla gardening often occurs when there are no legitimate alternatives for cleaning up blighted property and gaining temporary use of land. (See Figure 2.9.)

Figure 2.9. Guerrilla gardening can help beautify neglected plots of land.

The security of land tenure also influences the type of agricultural production and the range of suitable crops. For example, if land is available only for a finite period of time, an urban farmer may invest in fast-growing seasonal crops such as leafy greens and tomatoes instead of long-term perennial crops such as asparagus, rhubarb, and fruit or nut trees. Uncertain land tenure may also prevent or prohibit the use of sustainable, organic production practices. Composting—the creation of nutrient-rich soil from the decomposition of vegetable, fruit, and other plant material—is an inexpensive and sustainable method used to fertilize soil. However, composting requires a larger initial time investment than do conventional methods, such as the application of chemical fertilizers (Tixier and Bon 2006).

Despite these challenges, local governments, community land trusts, and urban agriculture practitioners are creating innovative ways to use land for agricultural purposes when lease agreements or ownership rights are unavailable. For example, a new policy in San Francisco allows developers to prevent their special permits from expiring if they allow urban agriculture on their sites during the period from permit approval to construction. Also, temporary or movable cultivation practices can be used to address time-related challenges. Contained growing systems such as raised beds, or temporary structures such as hoop houses, are easier to construct and require less time and investment than in-ground cultivation and greenhouse construction. The Resource Center, the nonprofit organization in Chicago that operates

Local governments, community land trusts, and urban agriculture practitioners are creating innovative ways to use land for agricultural purposes when lease agreements or ownership rights are unavailable.

City Farm, has an agreement with the city to use vacant, publicly owned land, with the understanding that it may need to relocate every couple of years. Despite the fact that City Farm has had to relocate four times in the past 25 years, it has been extremely successful, due in part to its founder's willingness to utilize flexible and temporary farming methods (Ward 2010). Garden State Urban Farms in Newark, New Jersey, utilizes a portable, raised-container farming system called Earth Box to grow a variety of fruits and vegetables. When Garden State Urban Farm's original site was slated for a new housing development, the portable farming system enabled quick and easy relocation to a new site. (See http://blog.grdodge.org/2009/09/28/have-farm-will-travel.) Such techniques allow urban farming to be an efficient interim land use for a variety of spaces.

In some communities, urban gardeners and farmers have developed contractual agreements with home owners to farm their backyard spaces in exchange for shares of the harvest. Backyard Urban Garden (BUG), a nonprofit urban farming organization in Salt Lake City, Utah, grows vegetables in three private backyards in exchange for weekly supplies of produce for property owners. BUG also sells produce in several farmers markets and through a CSA program. (See www.backyardurbangardens.com.)

In recent years, conservation groups and community land trusts have helped to create secure land tenure through ownership or long-term agreements (Caton Campbell and Salus 2003; Davis 2010, esp. part 5). For example, the Southside Community Land Trust (SCLT) in Providence, Rhode Island, works with residents to transform blighted vacant lots into community gardens by acquiring title to them and leasing the land to other organizations. Over the course of 25 years, SCLT has converted approximately five acres of formerly vacant lots into community gardens; expanded its farm operation to 50 preserved acres in Cranston, Rhode Island; established the Broad Street Farmers Market; developed a successful CSA program; grown, donated, and sold hundreds of pounds of organic produce; helped 15 schools start their own gardens and garden clubs; hosted young people at a children's garden; educated volunteers about urban environmental and local food issues; and assisted in the start-up of seven new minority-owned farm businesses. (See www.southsideclt.org/about.)

In Madison, Wisconsin, the Madison Area Community Land Trust (MACLT) holds title to the land on an unusual 31-acre development known as Troy Gardens, which encompasses community gardens, a working CSA farm, a restored prairie, an interpretive-trail system, and a 30-unit, mixed-income, green-built cohousing community. (See Figure 2.10.) Located on Madison's north side, the 26 acres of urban agriculture and open-space uses are managed by the nonprofit Community Ground-Works, which employs the farm manager, natural areas coordinator, education director, and other staff. (See www.communitygroundworks .org.) The Center for Resilient Cities, a conservation land trust and nonprofit resilience planning and design firm (www.resilientcities.org) holds a conservation easement on that same 26 acres and monitors site management (Caton Campbell and Salus 2003; Raja et al. 2008). The project moved forward through a planned unit development (PUD)

Figure 2.10. Troy Gardens, a community garden within a cohousing community protected by the Madison Area Community Land Trust

Alicia Acken

process that allowed for a "custom" zoning definition specific to the unique and complex mix of land uses on the site (Rosenberg 2010). This unusual partnership among two land trusts and a nonprofit that provides secure land tenure for a four-acre CSA farm and five acres of community gardens is now 12 years old.

To create more secure land tenure for urban gardeners and farmers, local governments can partner with community land trusts or other organizations. Planners can help by identifying and removing potential regulatory and policy barriers to the establishment of legitimate urban-agriculture activities as both long-term and interim land uses.

Location, Siting, and Land Use. Consideration of the location and siting of urban agriculture involves several important factors: competition with other public land uses; compatibility with neighboring land uses; location of nearby amenities; and agricultural intensity.

In many built-out cities, urban agriculture may compete for space with commercial development or public land uses such as parks, housing, schools, or other public infrastructure. Historically, urban agriculture has not been considered an appropriate use for public land, and uses that are not parts of plans or policies are often not acknowledged as important. Therefore, local governments should consider how urban agriculture can fit into larger urban systems, as well as redevelopment or revitalization efforts. For example, community gardens are appropriate in residential neighborhoods or in areas identified as food deserts, in school yards, or elsewhere throughout the city. Commercial beehives integrated into municipal green-roof programs can enhance pollination and plant growth. And establishing farmers markets at transit stops or other high-foot-traffic areas provides convenience for residents and a steady source of customers. Urban agriculture can also help integrate features such as wildlife habitat, stormwater management, or erosion control into designated areas (Mukherji and Morales 2010).

Urban agriculture can be integrated into private development as a permanent use (e.g., as part of open space requirements), but on its own it is usually not able to compete with other development opportunities that offer greater returns on investments. Therefore, in addition to innovative and flexible land-tenure agreements and flexible farming practices, careful siting of larger commercial and hybrid urban-agriculture operations is important to the success of urban agriculture. Higher-intensity agricultural activities such as large-scale plant or animal production, composting, food preparation and processing, or animal slaughtering may be better suited for commercial or former industrial areas on the urban periphery (Mukherji and Morales 2010).

Local government policies created to regulate land use in the public interest—such as zoning, subdivision ordinances, and design standards—may pose barriers. Urban agriculture is often not permitted as of right in residential, commercial, or mixed use zoning districts. Assigning urban agriculture to a single land-use category may be difficult due to its many different scales, intensities, and purposes. Further, local governments tend to view urban agriculture, especially community gardens, as an interim land use until a "higher and better" (i.e., tax-paying) use is identified. This creates additional challenges and uncertainties, particularly for publicly owned vacant land (Schukoske 2000). Planners are often not aware of the many benefits of urban agriculture and therefore may not factor it into redevelopment plans.

Municipalities that are largely built out must determine future uses for the remaining available parcels in the face of many competing public and private interests—a delicate balancing act that is not often appreciated by urban agriculture proponents. Public land therefore becomes an important resource for urban agriculture: the pressure to identify a higher and better use is not as acute as it is for privately owned parcels. This subject is treated in greater detail in Chapter 3 and the case studies included in Chapter 4. Considering these challenges, the ability to demonstrate and quantify the benefits—particularly the economic ones—of urban agriculture and develop innovative methods to reuse existing space may prove essential to the success of urban agriculture.

Natural Resources

The challenges of urban agriculture involve more than the accessing and securing of land; other resources are needed to make urban food production effective. Planners and other local government staff can develop and implement a variety of policies and programs and remove existing barriers to ensure that urban agriculture practitioners have access to healthy, uncontaminated soil, compost, and water.

Soil and Water Quality. Soil and water contamination are significant and often limiting factors for the reuse of urban sites for agricultural purposes. Such contamination can negatively affect plant growth and pose serious human health problems. A growing number of urban agriculture projects are established on brownfields, abandoned or underused sites where redevelopment or reuse is complicated by the presence of contaminants such as gasoline, diesel fuel, asbestos, heavy metals, solvents, lubricants, acids, and polychlorinated biphenyls (PCBs). For many, the term conjures images of large-scale industrial properties, but brownfields come in all shapes and sizes—from abandoned mining operations covering several square miles to vacant single-family homes with lead paint or asbestos insulation.

Direct sources of site contamination include the use of pesticides and fertilizers as well as chemical spills and leaks. Indirect sources may include stormwater runoff from contaminated surfaces or groundwater movement from adjacent polluted properties (Turner 2009). Mobile soil contaminants (such as petroleum, fuel oils, or dry-cleaning solvents) can percolate into groundwater, while other pollutants will remain on site, either weathering naturally or remaining unchanged in the soil. Prior agricultural use of land may also present a problem; for example, lead arsenic has historically been used for pest control in apple orchards, possibly resulting in elevated levels of lead in these soils (Shayler, McBride, and Harrison 2009).

Contaminated soil poses challenges for agricultural uses, as urban farmers, gardeners, and bystanders (particularly children) can absorb contaminants into their bodies via skin contact with, ingestion of, or inhalation of contaminated soil or plants (Turner 2009). How much of a contaminant is absorbed by plants and where within the plant the contaminant is stored—whether it remains in the roots or progresses to the shoots or fruits of the plant—is affected by soil, contaminant, and plant characteristics. The combination of these factors determines whether the contaminant will be passed on to people consuming the products (U.S. EPA 2010).

Considering the potential risks associated with reusing vacant or abandoned property for urban agriculture, requiring environmental site assessments (such as Phase 1 and 2 site assessments) is important.[2] After an assessment is complete, site-cleanup goals are developed according to the property's intended reuse plan (e.g., a housing development will have more stringent cleanup standards than a commercial development). However, while specific risk-based standards exist for residential, commercial, and industrial reuse of brownfield sites, they have not been tailored for urban agriculture reuse.

If contamination proves too cost-prohibitive to remedy, contained systems can be used to bypass exposure. These include both soil covers and contained food-production methods such as raised beds, hydroponic or aquaponic systems, and vertical or container-based gardening systems (Turner 2009). Together, soil covers and contained food-production methods reduce plant and human contact with contaminated soil. Such technologies are widely used throughout the United States and, depending on the system, can be low cost and low maintenance.

Unfortunately, many local governments do not require environmental site assessments, do not provide standards for safe and effective contained systems, and do not have standards for ensuring that imported soil and growing mediums, such as "clean fill," are safe and contaminant free. And recent research indicates that raised beds filled with fresh compost can become recontaminated over time, due to runoff and windborne dust from contaminated areas (Estes, Carter-Thomas, and Brabander 2010).

Given these challenges, planners can take a proactive role in encouraging local governments to (1) integrate provisions for environmental site assessments—particularly Phase 1 assessments, which provide a basic understanding of a site's history and past uses—in land-use regulations, vacant property management and foreclosure policies, site development and redevelopment policies, community garden and urban agriculture licensing programs, and other regulations; and (2) collaborate with state and regional EPA offices to develop recommendations for contained systems and clean fill.

Compost. The likelihood of contaminated soil on a site—or the lack of any soil structure at all—typically necessitates importing a growing medium from off site. Increasingly, serious urban agriculture operations try to establish a stream of organic compost inputs from nearby sources; these can include food waste

Figure 2.11. Compost piles are a common component of community gardens.

Kimberley Hodgson

from restaurants and grocery stores, leaves from municipal collection systems, coffee grounds from local roasters, and brewery waste from microbreweries. In many cases, worms are added to compost mixes, and through the process of vermicomposting their castings create a fertilizer extremely rich in nutrients.

Many urban agriculturalists produce as much of their own compost as possible, but additional demand can be met through either large-scale and centralized or smaller and decentralized composting operations. (See Figure 2.11.) Such operations may be restricted by zoning regulations. In Milwaukee, Growing Power and the Milwaukee Metropolitan Sewerage District (MMSD) are partnering to create large-scale composting operations on land owned by MMSD. The district has offered a long-term lease on a four-acre lagoon site for Growing Power to expand its composting operations. Once used to store sewage sludge, the site has not been used for this purpose in more than 15 years (Behm 2009). Other local governments have created municipal food-waste composting programs; examples include San Francisco; King County, Washington; Vancouver, British Columbia; and Toronto, Ontario.

Many opportunities exist to develop programmatic or policy actions to divert food-related wastes and other compostable materials from municipal solid-waste streams to publicly or privately owned composting operations. For more information about how local governments and planners can improve access to compost for urban agriculture practitioners, see Chapter 3.

Water Availability and Access. Water is required for plant growth. Without the availability of and access to water, urban agriculture will not be a successful community or economic development strategy. Municipalities can install infrastructure, establish policies and incentives, and provide affordable water permits to provide community gardeners and commercial urban farmers with access to water.

Few cities have urban agriculture–specific water policies on their books. One exception is the City of Cleveland, which provides community gardens with the option of purchasing seasonal permits allowing unmetered access to fire hydrants for irrigation; permits cost approximately $78. (See Cleveland-Cuyahoga County Food Policy Coalition n.d.) In order to provide affordable water access to new commercial urban garden and farm projects, the Cleveland Cuyahoga County Food Policy Coalition is working with the Cleveland Department of Water to develop new policies and water permits to address the growing urban agriculture needs within the jurisdiction.

Local governments, in partnership with U.S. Department of Agriculture (USDA) extension agents and other entities, can play important roles in determining water needs for varying sizes and intensities of urban agriculture and establishing water conservation standards to reduce overall irrigation allowances. Climate and the plant or animal species in question determine water requirements and influence the choice of irrigation system used. Overhead-irrigation systems, such as watering cans, water hoses, sprinklers, and perforated pipes are typically inexpensive and easy to use; however, they are inefficient, especially in drier climates, because some proportion of the water evaporates before it soaks into the soil. Above-ground drip- or trickle-irrigation systems use 10 to 20 percent less water but are more expensive than overhead-irrigation systems and require filters, pumps, and pressure regulators. Underground drip- or trickle-irrigation systems, also known as low-pressure directed-use systems, provide water directly to the soil and plant roots by capillary action. These systems are the most efficient and can decrease water usage by 50 percent (Pollock 2010). Further, the natural filtering effect of the soil can limit the transmission of pathogens from contaminated water to plants. Urban farmers can also employ soil-free techniques, such as hydroponics or "living biological systems," which

iStockphoto.com / Jon Schulte

Without the availability of and access to water, urban agriculture will not be a successful community or economic development strategy.

circulate water and nutrients through a closed-loop system combining fish and vegetable production.

Local governments can also develop urban agriculture production standards to minimize water loss through evaporation. Policies requiring the incorporation of soil amendments and organic materials into soil and the use of mulch or other material to cover the ground surrounding plants can help reduce water loss by 10 to 20 percent (Cleveland-Cuyahoga County Food Policy Coalition n.d.). Local governments can also explore alternative water sources such as rain barrels, cisterns, and other water collection and reuse systems. New York City's recently adopted stormwater plan calls for rainwater reclamation, or the diversion of culverts into rain barrels and cisterns; this could provide alternative irrigation sources for urban farmers. Each water source, however, presents different challenges ranging from cost to associated environmental and health risks and may require monitoring, depending on its quality (Tixier and Bon 2006; National Research Council 2010, 7). For example, rain barrels that divert water from rooftops may be contaminated with certain pollutants.

Operations

While planners have little control over the operations side of urban agriculture activities, they should still have a basic understanding of the marketing, human, financial, technical, and food-processing difficulties urban agriculture operations face in successfully scaling up their activities to fulfill their organizational or financial goals.

The volume of produce and proximity to consumers and markets, in combination with access to distribution and transportation infrastructure, greatly affects economic potential.

Scale, Volume, and Access to Market. In commercial urban agriculture, produce is grown for some degree of market sale. Crops are selected based on documented or speculative demand, and the operation is guided by a marketing plan. Urban markets include restaurants, farmers markets, schools and other institutions, and conventional or specialized retail outlets.

Kimberley Hodgson

The volume of produce and proximity to consumers and markets, in combination with access to distribution and transportation infrastructure, greatly affects economic potential. Transporting agricultural products from distant sites requires specialized delivery trucks and storage facilities. Such infrastructure is not typically needed to deliver urban agricultural products to local consumers, though urban growers may face extra transportation and labor costs if farming several noncontiguous plots.

Growers who wish to sell their products to schools, hospitals, or other institutions face additional challenges. In order to sell fresh food products during winter months, they may need to invest in commercial refrigerators or other cold-storage units. Depending on the size of the parcel under cultivation and the conditions of the lease, some growers may require additional on- or off-site storage space. Direct sales to institutions, especially schools, may require standardized processing practices as well as regular deliveries to multiple destinations. In addition, it can be difficult for urban agriculture practitioners to produce sufficient volumes to meet the needs of retail and institutional customers. Despite these challenges, there are many successful farm-to-institution programs across the country. Growing Power (Milwaukee) and Soil Borne Farm (Sacramento, California) are urban farms that sell to local schools and provide hands-on education activities for students and teachers.

Urban agriculture is a labor-intensive activity operating with few financial resources; urban farmers are frequently assisted by interns supported by grant funding or federal programs and dedicated volunteers.

Human, Financial, and Technical Resources. Urban agriculture is almost always a labor-intensive, nonautomated activity, operating with little in the way of financial resources. Many farmers operate their farms independently, though others are staff members of larger organizations whose missions are addressed through urban agriculture. The educational and social aspects of urban farms contribute to their operations; urban farmers are frequently assisted by interns supported by grant funding or federal programs such as Vista and AmeriCorps, and many depend on dedicated volunteers.

Kimberley Hodgson

Commercial urban farmers are often young and new to agriculture, and many acquire training through apprenticeships at established rural or periurban organic farms. That experience is then transferred to an urban operation and, in turn, taught to youth participants and volunteers. University extension offices in urban areas once offered technical assistance as well as master gardener programs. Though many of these offices are now reduced in size or nonexistent, where still active they can be at the forefront of urban gardening. In Philadelphia and Milwaukee, for example, the Penn State and University of Wisconsin extension offices, respectively, manage long-standing community gardening programs. Outside of Boston, University of Massachusetts extension agents offer valuable translation and marketing assistance to the state's many immigrant urban and periurban farmers. Still other entities, such as MOSES (Midwest Organic and Sustainable Education Service), an Upper Midwest nonprofit with a 10-year history serving a 12-state area (see www.mosesorganic.org), offer sustainable agriculture trainings, technical assistance with transition to organic production, farmer-to-farmer mentoring, free services linking land seekers with available land, and annual organic and sustainable farming conferences attended by urban as well as rural farmers. Increasingly, urban agriculture practitioners are finding educational value in trainings of this type.

Urban agriculture incurs costs for rent, liability insurance, labor, tools, equipment, water, transportation of inputs and outputs, marketing, and processing and packaging of products. Farms may require financial and technical assistance with site management and maintenance, education and training of volunteers and staff, and materials such as seeds, water hoses, gardening tools, perimeter fencing, storage bins and sheds, vehicles, signs, benches, walkways, and mulch (Kaufman and Bailkey 2000; Schukoske 2000; Brown and Carter 2003; Tixier and Bon 2006). In addition, special training may be needed on ecologically and economically sustainable production, processing, and marketing techniques; the potential health risks associated with the use of agrochemicals and untreated organic waste and wastewater; and proper hygiene in food processing and marketing activities (Dubbeling and Merzthal 2006). Commercial urban farms rarely generate enough revenues from sales to cover these costs and are often heavily dependent on other financial sources, such as foundation grants and donations.

Some urban farms, such as Growing Power in Milwaukee and Nuestras Raices in Holyoke, Massachusetts, have diversified their income streams through combinations of direct marketing to individuals, retail sales to restaurants and food co-ops, sales of value-added products, workshop offerings, training and technical assistance, and other fee-for-service endeavors. However, many urban farm projects lack a development staff; thus, fundraising and grant writing take up time that could be better spent farming. The federal government, through programs such as the USDA's Community Food Projects Competitive Grants Program, has been a supporter of urban agriculture. But such support is minuscule when compared with federal subsidies to mainstream agriculture (and commodity crops in particular; see sidebar, p. 34).

Figure 2.12. *Natural and value-added goods for sale at a farmers market in Virginia*

John Reinhardt

Processing and Added Value. Although most commercial urban agricultural products are distributed straight from the farm to consumers, some entrepreneurial operations look for ways to maximize revenue and social benefits through various processing and value-added enterprises. Such activities not only diversify the products of urban agriculture but also provide needed opportunities for learning how to begin and manage small food businesses. Typical value-added products include salad mixes, salad dressings, salsas, or honey from urban beehives, as well as nonedible products such as hanging flower baskets and packages of vermicompost. (See Figure 2.12.) The processing of urban-raised livestock such as poultry is possible but frequently limited by various state and local regulatory barriers.

Other Concerns

Security and Vandalism. Many urban agriculture activities are exposed to the public, making the vandalism of property, food, or equipment a po-

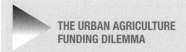

THE URBAN AGRICULTURE FUNDING DILEMMA

The consistent lack of funding poses a serious obstacle for the success of urban agriculture as a viable community development or economic strategy. While the USDA's Community Food Projects Competitive Grant Program provides funding for small-scale urban agriculture projects that address food insecurity in low-income communities, the amount allocated to this program is inadequate to cover the growing number and variety of urban agricultural projects throughout the country—and urban agriculture is not its specific program focus. The 2011 funding allocation is for a total of $5 million, with caps of $25,000 on Planning Grant applications and $300,000 over three years for Community Food Project Grants, compared to approximately $75 billion available in the commodities program under the 2008 farm bill (USDA ERS 2009).

Agriculture is still widely viewed as a rural, not urban, activity by many federal and state agencies. Despite opportunities to include urban agriculture activities in new and existing public housing, schools, and other civic spaces, the Department of Housing and Urban Development, the Environmental Protection Agency, and the Department of Health and Human Services offer little to no financial support. The USDA's National Institute for Food and Agriculture (formerly the Cooperative State Research, Education, and Extension Service), a nationwide, noncredit educational network of state land-grant universities and local or regional offices, provides "useful, practical, and research-based information to agricultural producers, small business owners, youth, consumers and others" primarily in rural areas (USDA NIFA 2008) but does not have the budget to support urban agriculture. Furthermore, few local governments provide financial resources to assist with urban agricultural start-up, management, and expansion costs (Kaufman and Bailkey 2000).

tential threat. However, this has typically not been a deterrent to urban agriculture. To minimize these threats, urban agriculture requires due diligence in securing tools and equipment. Many sites are fenced for protection from trespassers, but practitioners have learned that a sense of security can be maintained by having an open-gate policy by which any visitor—especially a neighbor—is welcomed onto the site and given a tour of the operation. In turn, and over time, the urban agriculture site will benefit from the well-documented "eyes on the street" phenomenon, where neighbors keep a watch on something considered a community asset (Nordahl 2009).

In cases where this is not enough, urban agriculture organizations can partner with local law enforcement and follow CPTED (crime prevention through environmental design) practices in the design of their garden or farm. Investing in insurance is another way to recoup the cost of stolen tools or damage from vandalism. However, there is often no substitute for building positive community relationships.

Two urban agriculture projects located in Milwaukee's Lindsay Heights neighborhood exemplify these principles and practices. Walnut Way Conservation Corp.'s unfenced peach orchards and production gardens are tended and watched over by organization staff and neighbors, who share in the harvest of produce. (See http://walnutway.org.) A few blocks away at Alice's Garden—a two-acre community garden bordered by a county park, an elementary school, and a number of vacant parcels awaiting residential redevelopment—program staff maintain an open-gate policy during daylight and early evening hours, inviting walk-in participation at cooking, gardening, and yoga classes and evening potlucks. (See www.resilientcities .org/?page_id=128.)

CONCLUSION

Urban agriculture addresses the traditional inputs, risks, and concerns of conventional small-scale agriculture (e.g., healthy soil and water, marketable crops, financial stability, etc.) as well as additional issues that arise in urban or suburban locations (e.g., land-use regulations, land access and tenure concerns, security and crime, etc.). Urban agriculture, due to its social, economic, and environmental benefits, should be considered part of a dynamic urban system that is understood by planners and influenced through the mechanics of planning practice. Chapter 3 develops in more detail the specific ways in which practicing planners can better facilitate urban agriculture through a variety of familiar tools and techniques.

ENDNOTES

1. The term "foodshed" was coined in Hedden 1929.

2. The first part of an environmental site assessment, Phase I, is an investigation of the potential for contamination based on the historic use of a property. If this assessment reveals a high probability of contamination, a Phase II environmental site assessment is necessary to confirm and evaluate the extent of contamination (Hersh et al. 2010).

CHAPTER 3

Facilitating Urban Agriculture Through Planning Practice

 As discussed in Chapter 2, urban agriculture is currently limited in its practice by a variety of overlapping challenges. Planners, if properly informed of these challenges, can play pivotal roles in removing barriers and supporting the integration of urban agriculture into the built environment and other systems.

John Reinhardt

Local governments can be supportive by providing citywide community-garden management or implementing municipal composting programs that make materials such as leaf litter available to the public.

This chapter describes the ways in which elements of planning practice—such as community engagement, data gathering and assessment, long-range goal setting, and policy formation and implementation—can help facilitate urban agriculture. Planners can help frame local urban agriculture practice within a larger physical and functional context. Local governments can be supportive by providing citywide community-garden management or implementing municipal composting programs that make materials such as leaf litter available to the public. A growing number of U.S. and Canadian municipalities are convening food policy councils, groups of public- and private-sector stakeholders that work to assess and strengthen their local food systems.

All urban agriculture operations function within local contexts and draw upon community resources. Land is the most important of these resources. Because land is limited and highly regulated in urban areas, planners must recognize urban agriculture as a land use in its own right to ensure its inclusion within the larger urban fabric. However, urban agriculture is less familiar to planners than traditional land-use categories and has a smaller constituency than other uses; as a result, planners have typically not been involved with local urban agriculture movements. They are instead seen by urban agriculture practitioners as responsive at best, as obstacles at worst, and most often as merely followers. No paradigm yet exists for the job of urban agriculture planner.

To help develop such a paradigm, this chapter describes a range of planning tools and practices that can help solidify urban agriculture as a standard concern for planners who act in the public interest. Through advocacy and the development of facilitative planning mechanisms and development regulations, planners can become champions of urban agriculture. This chapter also reviews a variety of local government actions related to urban agriculture that fall outside of traditional planning activities.

Like other forms of community development, planning for urban agriculture entails engaging with the market forces that affect urban and suburban neighborhoods, particularly disadvantaged ones. This chapter links the practice of urban agriculture to the five strategic points of planning intervention, as defined by APA's National Centers of Planning (www.planning.org/nationalcenters):

• long-range community visioning and goal setting,

• plan-making actions,

• standards, policies, and incentives to achieve desired plan goals,

• influencing the outcomes of development projects, and

• influencing the execution of public investment decisions.

DEVELOPING COMMUNITY VISIONS AND GOALS FOR URBAN AGRICULTURE

Ideally, the starting point for urban agriculture planning is the initiation of a community engagement process through which planners identify how urban agriculture contributes to the social, economic, and environmental goals of a community. A variety of approaches can be adopted. Those described below encompass: (1) governmental and nongovernmental partnerships, and (2) deliberative bodies, such as food policy councils.

Fostering Governmental and Nongovernmental Partnerships

Local governments, community development groups, and policy makers often question urban agriculture's benefits. Urban planners may have reservations about using high market-value land for urban agriculture. Community development groups sometimes doubt the ability of urban agriculture to create jobs and revitalize communities. Because the benefits of urban agriculture are not always easily quantifiable, community development groups often focus on other food-access strategies, such as siting new supermarkets in underserved urban neighborhoods. It is essential that groups familiar with urban agriculture coordinate and collaborate to educate different stakeholders and improve their understanding of urban agriculture's multiple benefits and social value (Kaufman and Bailkey 2000).

Collaboration among key community leaders is instrumental in the creation and implementation of effective, sustainable policies.

John Reinhardt

Collaboration among key community leaders is instrumental in the creation and implementation of effective, sustainable policies (Innes and Booher 1999). Unfortunately, fragmentation among professionals, policy makers, and the community is common, especially when addressing environmental issues that span several jurisdictions and fields of study, as urban agriculture does (Caton Campbell 2004). For example, public health, nutrition, and food professionals typically work together to improve specific food environments, such as schools and day-care facilities, but few local governments work collaboratively with these parties to improve the larger community food system (Feldstein 2007).

Planners should take advantage of this collaborative potential. Considering the variety of benefits urban agriculture brings to communities, agencies and organizations such as departments of planning, community development, parks and recreation, education, sanitation, aging, housing, public works, and others may prove to be useful allies in its support—especially those not historically involved with this use.

Through their understanding of how local government works and what it can allow, public-sector planners are well positioned to link local momentum for urban agriculture to appropriate facilitative mechanisms within government. They can also bring together urban agriculture practitioners and stakeholders both within and outside of government to identify common goals, the obstacles to achieving those goals, and the appropriate place of urban agriculture within other city systems (Caton Campbell 2004).

Food Charters

The articulation of a food charter—a broad vision statement and or set of guiding principles—often precedes a local or regional food-systems plan-making process. Developed through a community-participation process, charters typically establish a vision for the future of a municipality, county, or region's food system, often outlining specific values and principles related to various elements in that system, from production to disposal (Raja et al. 2008). Urban agriculture will typically play a role in the implementation of policies or programs that support a food charter. In the Vancouver (British Columbia) Food Charter, adopted by the city's mayor and city council, urban agriculture, in the form of edible gardening, is one of the charter's major goals. Appendix 1 provides information on food charters in additional municipalities.

Food Policy Councils

The potential for food policy councils (FPCs) to effectively facilitate public planning actions in support of urban agriculture has yet to be fully realized. As with other components of food-system planning, however, this, too, is steadily advancing, with urban-agriculture practitioners leading the creation of FPCs in metropolitan Kansas City, Milwaukee, Philadelphia, Chicago, Detroit, Calgary, and elsewhere. In a 2009 nationwide study, Food First and the Community Food Security Coalition noted that FPCs generally assume four functions (Harper et al. 2009):

Food policy councils take a variety of forms and can operate at several scales: municipal, county, state, or region.

(1) to serve as forums for discussing food issues;

(2) to foster coordination between sectors in the food system;

(3) to evaluate and influence policy; and

(4) to launch or support programs and services that address local needs.

FPCs view all components of local or regional food systems as linked. A local urban agriculture strategy, for example, will likely be considered a means to achieve broader goals, such as improving modes of access to fresh, affordable food. Through systemic examinations of food issues and the formulation of recommendations, FPCs act as food-system integrators that can champion relevant programmatic and policy actions by elected officials and planners.

FPCs take a variety of forms and can operate at several scales: municipal, county, state, or region. They can be formally created by direct government action, such as a resolution, or work independently of government; most local and county-level FPCs are the latter type (Harper et al. 2009). In either case, FPCs must gain some legitimacy from government to maintain their potentials to affect food policy. If an FPC is a formal arm of government, this better justifies the direct participation of government representatives in the FPC's work, though it may also constrain the council's scope or recommendations. The bimonthly meetings of the Toronto Food Policy Council, for example, a unit of that city's Board of Health, regularly draw agency representatives who have some stake in a particular agenda item. Another pioneering FPC, the Portland/Multnomah Food Policy Council in Oregon is an advisory body to the city's Bureau of Planning and Sustainability. Composed of community representatives, it is staffed through the bureau's Sustainable Food Program, thus establishing a functional connection to city government (Raja et al. 2008, 68–76).

There is no established template for how FPCs should work with governments or quasi-governmental agencies in the cause of urban agriculture. Because the sociopolitical and community food-system environments within which FPCs form are quite variable, such collaborations tend to be circumstantial, opportunistic, and improvisational. In western Michigan, for example, the Greater Grand Rapids Food Systems Council, a nongovernmental FPC, engaged the planning staff of the Grand Valley Metropolitan Council, the regional council of governments, in exploring the possibilities of incorporating urban agriculture into the Grand Rapids zoning code. Their approach employed form-based zoning to envision the physical characteristics of urban agriculture when practiced in three urban "transects" with different residential densities. And in 2009, members of the Dane County (Wisconsin) Food Council, a formal committee of county government, seized the opportunity of a complete rewriting of Madison's zoning ordinance to assist city planning staff in rethinking community gardens and urban agriculture. As a result, Madison has created a new Special Urban Agriculture District permitting community gardens and for-market urban farms within most zoning districts. The city also looked at the existing periurban agriculture zoning district and revised it to better establish larger-scale urban agriculture within expected development on the city's fringe. In the absence of a city-level FPC in Madison, the food council's standing within Dane County government gave it the legitimacy to consult with Madison's city planners in this cross-jurisdictional collaboration.

PLAN MAKING

Local and regional governments—and their planning agencies—play important roles in legitimizing urban agriculture as a recognized land use or community development strategy. By identifying existing community needs that urban agriculture can address, inventorying necessary local resources, and evaluating current policies and legislation, local governments can work to effectively integrate urban agriculture considerations into the plan-making process.

A key indicator of the legitimization of urban agriculture as a planning issue is its increasing appearance in comprehensive, strategic, functional, and subarea plans, as well as in plan implementation mechanisms such as zoning. In each case, urban agriculture is deemed important enough to the public interest to have a part in the long-term future vision outlined in a plan and in the programs and policies used to implement that vision.

Documenting Existing Conditions: Food-System Assessments and Resource Surveys

As part of municipal and regional plan-making processes, planners typically identify, document, and analyze the social, economic, and environmental characteristics of a community. While not within the traditional domain of planning professionals, the food-system assessment (and its individual parts) is an important tool that can be used to study food production, processing, distribution, retail, access, consumption, and waste within a community. Topics covered by such an assessment may include:

• stakeholders

• socioeconomic and health statistics

• household food security

- culturally appropriate food production, processing, distribution resources, trends, and economic activity

- land availability and suitability for food-system activities such as urban agriculture

- location and number of food sources and outlets within a community

- availability, affordability, and nutritional quality of foods sold in these outlets

- existing governmental and nongovernmental programs and policies.

In an effort to better understand the regional food system in the greater Philadelphia area, the Delaware Valley Regional Planning Commission conducted one of the first regional food-system studies in the country, *The Greater Philadelphia Food System Study* (DVRPC 2010). The study evaluated agriculture resources, food-production trends, natural-resources constraints, the economic significance of the food economy, public health, and consumption patterns, among other characteristics of the greater Philadelphia food system. (See Figure 3.1.) It identified opportunities and problems within regional and local food systems and serves as a foundation for future long-range goal setting, policy development, and implementation measures to address identified problems.

Community Food Assessments. The community food assessment (CFA) is a specific type of food-system assessment, typically focusing on food-security indicators within a defined area. Developed in the 1990s, the CFA is defined as "a collaborative and participatory process that systematically examines a broad range of community food issues and assets, so as to inform change actions to make the community more food secure" (Pothukuchi et al. 2002).

A CFA thus combines aspects of rational planning (the systematic analysis of a problem and the formulation of solutions) and participatory planning (the active inclusion of a range of stakeholders) to formulate a set of interventions built on identified community food issues and assets. Typically, the CFA will include urban agriculture strategies within its set of recommended actions.

CFAs are conducted at a variety of scales from neighborhood to region; the complexity of the assessment depends on the size of the area being studied. A CFA can take a year or longer to produce and is often conducted by a combination of experienced food-system analysts and community stakeholders. Both quantitative and qualitative methods of data gathering are typically employed.

One model CFA is the Vancouver Food System Assessment of 2005 (www .sfu.ca/cscd/research-projects/food-security). Written by a team of local food-system researchers with the support of the city's Department of Social Planning, the Vancouver CFA is a comprehensive scan of food-system needs across all 23 city neighborhoods. It gauges overall levels of food access, promotes strategies to improve food security citywide, and is intended to inform the work of relevant city agencies and the Vancouver Food Policy Council. The CFA's recommendations seek to create an alternative, food-related social economy by using relationships among food-system actors (such as networks of small businesses) to develop a supportive infrastructure for food enterprises (e.g., product development, training, and marketing) and to better establish an economic foundation for urban agriculture. Specific recommendations include expanding urban agriculture as a component of city-led development, as well as increasing the number of community gardens citywide (as in Figure 3.2).

Figure 3.1 *The Greater Philadelphia Food System Study was one of the first regional food-system studies in the country.*

Figure 3.2. Food-systems planning has taken on increasing importance on many scales in Richmond, near Vancouver, British Columbia.

Comprehensive Urban-Agriculture Studies. Forward-thinking studies, both within and outside of government, have articulated how urban agriculture can be expanded to meet a metropolitan region's social, environmental, and public-health goals. For example, the Planning and Housing Committee of the Greater London Authority in England authored a 2010 report that challenged the mayor of London to better incorporate urban agriculture into a number of existing planning actions within the context of the broader London Plan. The report's recommendations included policies promoting the incorporation of urban agriculture into the Local Development Frameworks of individual London boroughs, integrating urban agriculture within the London Plan's waste, water, and energy policies, and specifying urban agriculture within the London Plan as a beneficial land use within the Greater London Green Belt (Greater London Authority 2010).

Also in 2010, Toronto's Metcalf Foundation released a report, coauthored by MetroAg: Alliance for Urban Agriculture, that proposed ways that urban agriculture in greater Toronto—present now (Figure 3.3), but not significantly shaping local food supply—can be scaled up throughout the metropolitan region. The report calls for public and private actions in five areas: (1) increasing urban growers' access to growing spaces; (2) creating a better physical infrastructure for urban agriculture; (3) strengthening the local food-supply chain; (4) building a knowledge infrastructure among urban agriculture

Figure 3.3. Urban agriculture is present in Toronto, as shown by the Wychwood Barns development, but a recent report calls for scaling up urban agriculture activities within the city.

practitioners and stakeholders; and (5) creating new governance, coordination, and financial support models (Nasr, McRae, and Kuhns 2010). As in the London study, the recommendations of the Toronto study build upon a comprehensive scan of current policies and programs and an analysis of whether they serve the goal of increasing urban agriculture.

Studies of Land Resources. As part of a larger food-system assessment, *The Greater Philadelphia Food System Study* examined the resource potential of periurban agricultural land in the Philadelphia region as the source for a more localized food system. Its scope of land-use analysis was the "100-Mile Foodshed": 70 counties across five states within a 100-mile radius of Center City Philadelphia—an area including New York City and Baltimore that has a population of more than 30 million people—that represent the production area for Philadelphia's regional food system (Figure 3.4). The DVRPC study found that 27 percent of the region's land is in agricultural use (with 16.9 percent having prime farmland soils) and that the region's farms are both increasing in number and growing smaller in size. Farming traditions continue in southern New Jersey and southeastern Pennsylvania in the face of increasing development, and those farmers supply food to nearby urban markets. However, the study determined that the Philadelphia region's growing population, and the resulting need for developed land, will lead to a significant deficit in the amount of agricultural land necessary to meet the demand for locally grown food. This finding suggests a limit to periurban agriculture as an alternative to transporting food over long distances.

Figure 3.4. Aspects of Philadelphia's 100-mile foodshed

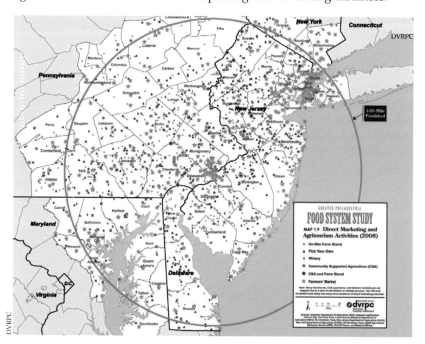

The City of Minneapolis has recently undertaken an intensive screening process of land that might be appropriate for urban agriculture, with input from the departments of housing, economic development, and planning. The result has been a list of nonbuildable lots determined to be good community-garden sites based upon a variety of criteria, such as access to sun and water and degree of soil contamination (City of Minneapolis 2010). The city's goal is to provide long-term if not permanent access to land for community gardens.

Two academic studies have used public-land inventories to determine land availability for urban agriculture and structure larger advocacy approaches for its expansion. In 2009, Nathan McClintock and Jenny Cooper of the Department of Geography at the University of California, Berkeley, produced a comprehensive analysis of the potential for urban agriculture on publicly owned land in nearby Oakland. Using a variety of tools, including GIS mapping, McClintock and Cooper (2009) determined that 495 sites totaling 1,200 acres could conservatively produce 5 to 10 percent of Oakland's vegetable needs. (See Figure 3.5.)

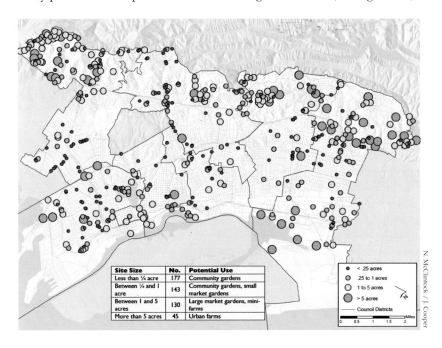

Figure 3.5. Cultivating the Commons inventoried 495 sites within Oakland, California, and analyzed their urban agriculture potential.

Site Size	No.	Potential Use
Less than ¼ acre	177	Community gardens
Between ¼ and 1 acre	143	Community gardens, small market gardens
Between 1 and 5 acres	130	Large market gardens, mini-farms
More than 5 acres	45	Urban farms

In November 2004, the City of Portland passed Resolution 36272, which initiated an inventory of city-owned land and its suitability for community gardens and other urban agricultural uses. A team of graduate students from Portland State University's Nohad Toulan School of Urban Studies and Planning collaborated with the Portland Water Bureau, Portland Parks and Recreation, the Bureau of Environmental Services, and the Office of Transportation to complete an inventory of vacant, publicly owned land; to analyze the barriers and challenges to urban agriculture in the city; and

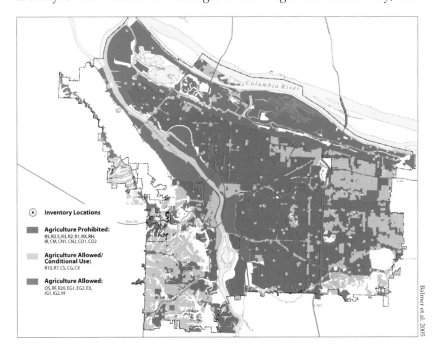

Figure 3.6. Portland State University graduate students identified 289 sites within the city as suitable for urban agriculture activites.

to make recommendations to the city on ways to incorporate urban agriculture into planning practice. Their 2005 report, *The Diggable City: Making Urban Agriculture a Planning Priority*, identified 289 of 875 potential sites as suitable for urban agriculture (Balmer et al. 2005). In response to the report, the Portland City Council directed the Portland/Multnomah County Food Policy Council to advise it on the barriers to urban agricultural uses on city-owned properties and on an appropriate management plan for these lands. The FPC established an urban agriculture technical advisory committee to research best practices and provide recommendations on land-use planning and zoning, immigrant farmer programs, community-supported agriculture, community gardens, public involvement, and other urban agricultural issues. In February 2006, the FPC's final report recommended that the City Council (1) focus on additional filtering of the inventory for its urban agriculture potential, (2) create pilot projects, (3) develop a land-management plan, and (4) explore policy changes to remove barriers. That fall, the Office of Sustainable Development received a grant from the USDA Risk Management Agency to complete the recommendations.

The Minneapolis, Oakland, and Portland land inventories explored the extent of possible food-production sites within city boundaries. Other tools can be employed by planners to examine the conditions for urban agriculture. In Indianapolis, geographic information systems (GIS) mapping has been used to identify spatial correlations between vacant parcels that could be used for urban agriculture and the presence of "food deserts," which are areas that lack full-service food retailers. Using GIS mapping, Cleveland has identified distances between households and community gardens. Few cities, however, have integrated brownfield assessments with urban agriculture land inventories.

Local Comprehensive Plans

While not all local governments across the United States are required by state statute to develop comprehensive plans, many that are are beginning to see the connections among comprehensive planning, neighborhood development and revitalization, health, food policy, and sustainability. The comprehensive planning process can be used to identify local economic, social, and health issues; engage and educate the community; and promote the long-term health of the community (Kelly and Becker 2000; Stair, Wooten, and Raimi 2008). Typically, when urban agriculture is considered in the comprehensive planning process, it is viewed as a strategy to achieve larger social or environmental goals—not as an end in itself.

Open-space goals and policies can encourage the conversion of vacant or abandoned land to urban agriculture and the preservation of existing urban agriculture. Economic development goals and policies can lead to new financing tools for urban agriculture development: tax incentives can encourage the location of urban agriculture in underserved neighborhoods on vacant property; other incentives can encourage public institutions to sell or use locally produced foods; business-enhancement incentives can encourage partnerships between food outlets and neighborhood-based nonprofits to encourage stores to offer locally produced foods; and public financing for private infrastructure can help improve the refrigeration or warehouse capacity of urban farmers to successfully sell perishable foods (e.g., fruits and vegetables). Housing goals and policies can encourage urban agriculture near affordable housing through the provision of community gardens, rooftop gardens, and community kitchens in multifamily and low-income housing projects (Center for Civic Partnerships 2003; Caton Campbell 2004; Flournoy and Treuhaft 2005; McCann 2006; APA 2007; Ashe, Feldstein et al. 2007; Feldstein 2007).

In its new general plan, the City of Richmond, California, acknowledges the benefits of urban agriculture, such as serving as a viable economic development vehicle. However, urban agriculture in the Richmond general plan is primarily seen as a strategy to reach certain community health and wellness goals (Figure 3.7). Under the goal to "Expand Healthy Food and Nutrition Choices," one specific policy recommendation encourages the city to promote urban agriculture on suitable publicly owned vacant sites; this is in addition to promoting the greater availability of fresh fruits and vegetables in Richmond's retail outlets and encouraging the city's restaurants to serve healthier fare (Richmond 2009, 11.38).

Figure 3.7. Urban agriculture is featured as an aspect of community of health and wellness in the general plan of Richmond, California.

Madison's 2006 comprehensive plan recognizes that the physical growth it guides will, to some degree, come at the expense of the high-quality farmland surrounding the city, as future annexation will bring working agricultural land within city boundaries. However, food production is seen as a vital activity within Madison city limits, and the plan supports the continuance of small-scale farming for urban markets. The larger food-system goals of the comprehensive plan are to be furthered by identifying existing farming operations within city boundaries, providing incentive programs for new and current farmers, encouraging a variety of agricultural uses (such as orchards) that are "compatible with urban uses," and protecting existing community gardens while facilitating new ones (Madison 2006, 2: Objectives 11–14; Raja et al. 2008, 47–48). Madison's comprehensive plan suggested a revision of the city's zoning code, and (as described above) the inclusion of urban agriculture four years later in the zoning code rewrite far exceeded what the plan envisioned. City planning staff attribute this to emerging trends and grassroots support since 2006.

As the city of New Orleans rebuilds from Hurricane Katrina's damage, its new master plan, approved in August 2010, assumes extraordinary significance, especially since, for the first time, its recommendations have the force of law (Eggler 2010). Urban agriculture appears in several of the plan's sections within the context of sustainable growth and development. The New Orleans case is discussed in detail in Chapter 4.

Many additional local governments are beginning to explicitly address urban agriculture in their local comprehensive plans. For an overview of some of these local governments, see Appendix 2.

Municipal Sustainability Plans

Given today's concerns for the energy use and resulting greenhouse gas emissions in metropolitan systems, North American cities are creating strategic frameworks to guide the adoption and implementation of sustainable development practices. The relatively new concept of community food systems lends itself to inclusion in these sustainability plans.

The environmental threats caused by greenhouse gases and global warming have led some municipalities to include urban and periurban agriculture as a "shadow" recommended strategy in their planning documents.

Kimberley Hodgson

The environmental threats caused by greenhouse gases and global warming have led some municipalities to include urban and periurban agriculture as a "shadow" recommended strategy in their planning documents—mentioned but not detailed. One such example is the Climate Protection Plan adopted by the Kansas City, Missouri, City Council in 2008. The plan recommends the promotion of neighborhood food production as a carbon-offset and waste-management strategy. Yet unlike other recommended strategies such as green roof development, the plan does not specify the amount of greenhouse gas emissions that would be reduced citywide through neighborhood gardens, reflecting a current shortage of available and relevant research. In contrast, the City of Toronto took a more explicit approach, specifically recommending localized food production in its 2007 Climate Change, Clean Air and Sustainable Energy Action Plan. Two of its implementation recommendations—an interagency urban-agriculture working group and the neighborhood-focused Live Green Toronto program—are now at the forefront of city government's support for urban agriculture.

A model example of linking urban agriculture to overall environmental strategies can be found in Baltimore's 2009 Sustainability Plan, which incorporates urban agriculture into two related approaches (Baltimore Office of Sustainability 2009). First, its "greening" theme includes the specific goal to "establish Baltimore as a leader in sustainable, local food systems," with the accompanying strategy of "increas[ing] the percentage of land under cultivation for agricultural purposes." The proposed policy language reads:

Increase the amount of food production within Baltimore City through a variety of approaches. Modify zoning regulations to accommodate urban agricultural production and sales. Increase the number of City farms and gardens in parks, on vacant lots, school grounds, and other appropriate and available areas. Promote community gardening for food production through programs such as the existing Master Gardener Urban Agriculture Program. Lastly, develop incentives and support for urban farm enterprises. (74)

Perhaps more noteworthy is the follow-up strategy to create a citywide urban agriculture plan:

Develop a plan that will promote healthy, local, and, where possible, organic food production and food professions, and include multiple stakeholders currently involved in food production and job training. The plan should identify the predicted demand for urban farmed food and recommend location and distribution of urban agricultural institutions. It could also identify the best distribution of existing food networks and identify gaps that need to be filled. (75)

Elsewhere in the Baltimore sustainability plan, under the "Cleanliness" theme, urban agriculture, as a form of community-initiated use of open space, is implicitly offered as a vehicle to "transform vacant lots from liabilities to assets that can provide social and environmental benefits" (35). This particular strategy recognizes that Baltimore, like other postindustrial cities, has a significant number of vacant properties (close to 30,000) and that facilitating the stewardship of its land resources is in the city's long-term interests.

Appendix 3 provides a brief overview of other local sustainability plans and how they address urban agriculture.

Regional Plans

Regional plans, which by their nature cover several jurisdictions and address functional linkages among them, lend themselves well to food-systems planning. For example, a regional plan can identify working agricultural land on the metropolitan fringe as a necessary resource to maintain the flow of locally grown produce to urban and suburban consumers. The scale of regional plans that cover multiple counties may, however, limit the overall attention paid to urban agriculture within those plans. (See Appendix 4.)

In 2003, the region of Waterloo in southern Ontario, which includes the cities of Waterloo, Kitchener, and Cambridge—one of the fastest-growing areas in Canada—adopted a Regional Growth Management Strategy to guide and control its growth and the resulting social costs. The strategy spurred the initiation of 80 "implementation projects" across the region's different agencies. The public health department used the opportunity to explore connections between the built environment and public health. Using an extensive, research-based approach, it produced *A Healthy Community Food System Plan for Waterloo Region* in 2007 (Figure 3.8). The far-reaching plan is founded on the belief that a regional health agency can develop a strong collaborative relationship with regional planners. The plan's treatment of urban agriculture focuses on backyard, community, and rooftop gardens as ways to achieve the objective: "Increase availability of healthy food so that healthy choices are easier to make" (Region of Waterloo Public Health 2007, 13).

The Delaware Valley Regional Planning Commission (DVRPC), author of the 2010 Philadelphia region's agricultural land-resource survey (above), provided the context for that survey in its 2009 regional plan, *Connections: The Regional Plan for a Sustainable Future*, a 25-year plan for the nine-county Philadelphia region (DVRPC 2009). The plan is structured on four "key principles," with two of the four—"Manage Growth and Protect Resources" and "Develop Livable Communities"—making reference to urban agriculture.

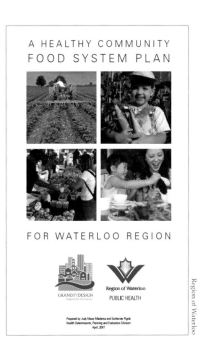

Figure 3.8. In Ontario, the Region of Waterloo's public health department created a food-systems plan intended to increase the availability of healthy food within the region.

The growth-management strategies include the localization of regional food production through increased periurban agriculture; only a passing reference to urban food production is made. Urban agriculture is advocated for more strongly under the "Develop Livable Communities" principle, as a component of a new green infrastructure across the region.

Given the wide extent of regional concerns addressed in the *Connections* document, the minor presence of urban agriculture is understandable. However, future regional plans—as well as more localized plans—can better highlight the value of urban agriculture and specify the dimensions of its practice, once plan authors become more familiar with its characteristics. Currently, the value and benefits of urban agriculture in a particular location is largely understood and conveyed anecdotally, unlike formal research on other food-system elements such as retail food access.

IMPLEMENTATION MECHANISMS FOR DESIRED PLAN GOALS

Although food systems and urban agriculture are gradually finding their ways into planning documents, neither will be fully codified without their accompanying inclusion in mechanisms executing the vision and goals of the plan documents. The regulatory nature of plan-implementation mechanisms, and planners' relative lack of familiarity with urban agriculture, means that developing appropriate implementation and regulatory language for the practice of urban agriculture is a significant policy issue for urban-agriculture advocates.

Zoning for Urban Agriculture

Although zoning is a common tool used by local governments to exercise their police power in the interest of public health, safety, and welfare, few governments have used zoning to improve the food environment. Urban agriculture is generally not permitted as of right in residential, commercial, or mixed use zoning districts, as discussed in Chapter 2. Urban agriculture, in particular the community garden, is often considered an "interim" use of land, creating additional challenges to and uncertainty about its permanence (Schukoske 2000).

While many suburban and exurban cities have agricultural zoning districts permitting a wide range of agricultural activities, relatively few communities explicitly acknowledge urban agriculture in their zoning regulations. Municipalities have traditionally viewed agriculture lands as "holding zones" for future development. Agricultural district designations are typically remnants of a rural past, applying to the hinterlands surrounding cities. Furthermore, while the large minimum-lot sizes and broad allowances for intense crop cultivation and animal husbandry found in typical agricultural district regulations are compatible with rural areas, such elements make these regulations inappropriate for denser city neighborhoods.

In some built-out cities, zoning references to agriculture are completely absent. In others, basic crop production may be permitted as an accessory use in low-density residential districts. While most zoning ordinances in urban areas do not preclude backyard gardening for personal use, on-site sales and animal husbandry are strictly limited or prohibited altogether.

This absence of urban agriculture within municipal zoning codes is starting to change, however. Communities looking to sanction or encourage urban agricultural activities on private land have taken a range of approaches; the most common is to list certain urban agricultural activities as permitted uses in existing zoning districts. Alternately, some jurisdictions have created new zoning districts to set aside specific areas for community gardens or urban farms. Some communities have also included urban agriculture as a desirable amenity within planned unit development (PUD) or traditional neighborhood development (TND) project guidelines or in conservation subdivision regulations. For example, the 31-acre Troy Gardens project in Madison—which combines an urban farm, community gardens, and affordable housing—was approved by the city as a PUD (Raja et al. 2008, 43–46).

Existing Zoning Districts. When communities choose to allow urban agriculture as a permitted use in existing districts, they may treat specific activities such as crop production and animal husbandry as distinct uses. Greensboro, North Carolina, and Toledo, Ohio, illustrate this approach by permitting crop production by right in most zoning districts (Greensboro 2010; Toledo 2010), while Cleveland and Seattle allow poultry, livestock, and bees in a wide range of residential and nonresidential districts subject to specific development standards (Cleveland 2010, sec. 347.02; Seattle 2010a). In addition, a number of communities have recently added community gardens as a permitted land use in certain districts. Definitions and use standards for community gardens often explicitly distinguish collective gardening efforts from commercial agriculture by prohibiting on-site sales. For example, Safety Harbor, Florida, defines *community garden* as follows:

> An area of land managed and maintained by a group of individuals or non-profit organization to grow and harvest food crops and/or non-food, ornamental crops, such as flowers, for personal or group use, consumption, donation, or non-profit sale at a location off-site from where the community garden is located. (Safety Harbor 2010)

A few jurisdictions have gone beyond community garden allowances to create special-use categories for more intense urban-agriculture efforts. Austin, Texas, for example, permits one- to five-acre urban farms with on-site sales in a variety of residential and nonresidential districts (Austin 2010). Similarly, Philadelphia, Denver, and Seattle have all recently proposed zoning amendments that would allow commercial agriculture in many existing districts (Philadelphia 2010b; Denver 2010; Seattle 2010b).

New Zoning Districts. As an alternative (or supplement) to adding urban agricultural activities as permitted uses to existing districts, some communities have created new zoning designations to recognize the importance of dedicated productive lands in urban environments. These new districts may protect or encourage community gardens or small-scale

Some communities have created new zoning designations to recognize the importance of dedicated productive lands in urban environments.

John Reinhardt

commercial farms. Boston, for example, allows community gardens to be rezoned as open-space subdistricts (Boston 2009), while a number of other cities have codified new districts to accommodate crop production, animal husbandry, and on-site sales. Some of these, such as the urban agricultural zone in Chattanooga, Tennessee, require large lot sizes in anticipation of a wide variety of intense agricultural uses (Chattanooga 2009). Others, such as Cleveland's Urban Garden district, allow urban farms on standard city lots (Cleveland 2010, chap. 336). For examples of additional urban agriculture zoning regulations, see Appendix 5.

Other Local Policies and Regulations Supporting Urban Agriculture

A host of other local policies besides zoning can be used to sanction or encourage specific activities related to urban agriculture. These policies can be grouped into four basic categories:

- Nonzoning regulations that affect the use of private land for agricultural activities;

- Land-use policies that permit public land to be used for gardens or farms;

- Land-disposition policies that permit surplus properties to be acquired for urban agriculture; and

- Policies and regulations that strengthen the infrastructure for widespread urban agriculture.

Nonzoning Regulations. Two common nonzoning regulations that communities can use to encourage urban agriculture on private land are animal control ordinances and residential composting ordinances. Over the past several years, many cities have revised their animal control ordinances to allow backyard chickens, livestock, and bees in residential districts (Figure 3.9). San Antonio, Texas, for example, permits three chickens and two larger farm animals in all low-density residential districts (San Antonio 2010). In addition, an increase in the number of urban dwellers interested in sustainable gardening has led to increased backyard composting. Because improper composting techniques can cause odors and attract vermin, a number of jurisdictions have codified standards for residential composting. Chicago exempts residential

Figure 3.9. Some zoning codes and animal control ordinances permit poultry, livestock, and bees in urban neighborhoods.

Nevin Cohen

composting from permitting requirements as long as the composting complies with a list of simple standards (Chicago 2010). For additional examples of municipal animal-control ordinances and residential composting ordinances, see Appendixes 6 and 7.

Public Land-Use Policies. Because not all urban agriculture occurs on private land, some communities have adopted policies that sanction the use of certain public lands for food production. These policies may create a mechanism whereby surplus public land can be used for urban agriculture as an interim use. Alternately, these policies may authorize residents to appropriate specific types of public land for garden space. In Hartford, Connecticut, the city code directs the parks and recreation advisory committee to maintain an inventory of vacant public lands, to establish a procedure for matching gardeners to available lots, and to adopt use standards for community gardens on public lands (Hartford 2010). Similarly, Escondido, California, has established an adopt-a-lot program to facilitate temporary-use agreements between citizens or neighborhood groups interested in creating community gardens and either the city or private landowners with vacant land (Escondido n.d.).

Apart from surplus public properties, other public land such as vegetated buffers between sidewalks and roadways or utility corridors may also be transformed into linear garden spaces. In 2009, in response to citizens' complaints, Seattle's Department of Transportation eliminated the $250 permitting fee for residents who want to grow food in right-of-way planting strips between the sidewalk and the roadway (Figure 3.10; Seattle DOT 2009).

Land-Disposition Policies. In communities with an abundance of surplus public properties, a clear land-disposition policy can be an effective tool for transferring underutilized sites to food producers. Vacant municipal land earmarked for future residential, commercial, or institutional uses or deemed not fit for con-

Seattle Department of Transportation

Figure 3.10. Seattle no longer charges permitting fees to residents who grow food in roadway right-of-way planting strips.

struction (areas such as flood zones, buffer zones, utility rights-of-way, land under power lines, and institutional property) may be suited for temporary or even permanent urban agricultural use. Both governmental and nongovernmental agencies and organizations can play active roles in providing temporary or long-term leases for this land (Brown and Carter 2003; de Zeeuw, Dubbeling, et al. 2007).

Local governments can also develop ordinances that permit the assigning of vacant municipal land under contract to urban agriculture groups for farming purposes. These permits could incorporate minimum parcel requirements for farming purposes (e.g., access to water). Furthermore, local governments can provide property-tax exemptions for community garden organizations and urban agriculture groups seeking to obtain ownership of a vacant parcel of land; establish usufruct agreements that permit the legal right to use public or private land in return for maintenance and upkeep of the land; or help negotiate tenure agreements between urban growers and private or public landowners with unused areas such as hospital grounds, school yards, university campuses, church grounds, and business rooftops (Mougeot 1999; Kaufman and Bailkey 2000; Schukoske 2000).

Vacant land is often privately owned, but local governments and nongovernmental organizations can work together to promote its use for urban agriculture. They can bring urban growers in contact with landowners to negotiate long-term leases, or they can lease land from private owners and then sublease it to community groups for urban agriculture use. In addition, local governments can provide tax incentives to landowners who make idle parcels available for urban agricultural use, or, conversely, they can increase municipal taxes on such parcels. They can also provide exemptions from municipal water fees (de Zeeuw, Dubbeling, et al. 2007).

Land banking is an important vacant-land management strategy. State, county, and municipal governments have created land-bank authorities for the explicit purpose of taking title to property, holding the property, and conveying it to others. By centralizing acquisition of vacant property into one agency, local governments avoid cumbersome legal processes and difficulty in clearing titles (Mallach 2006). For urban agriculture practitioners, land banks such as the Genesee County Land Bank in Flint, Michigan, can serve as clear avenues to the acquisition of land parcels, often through mechanisms such as side-lot sales to adjacent landowners. Although land banks are a valuable land-conveyance mechanism, they do entail risks, including uncertain environmental conditions and market values. In addition, land banks may not have disposition strategies consistent with guiding comprehensive land-use plans (Federal Reserve Bank of Atlanta 2009).

Conservation easements and land trusts can also play important roles. In much the same way that rural agricultural land is preserved through conservation easements, urban land may also be protected for agricultural use. For example, Minneapolis's Real Estate Disposition Policy explicitly permits nonprofit organizations and other public agencies to purchase surplus lots for use as community gardens. Purchasers exercising this option must place a conservation easement on the lots to ensure that all future owners will use the space for gardening.

While easements are useful tools for preserving agricultural land, not all states have adopted enabling legislation. Acquiring and preserving urban land for agriculture may prove to be much more difficult than for rural land because of a variety of development pressures and competition from other land uses. A land trust can negotiate conservation easements with private landowners, though it is often difficult to locate landowners of abandoned or tax-delinquent properties (Brown and Carter 2003). While easements represent a valuable mechanism, community and conservation land trusts can practice broader strategic approaches to supporting urban and periurban agriculture. A land trust may be particularly helpful if it concentrates less on pursuing individual easements and more on purchasing clusters of periurban farms or strategically preserving threatened or culturally important community gardens.

For additional examples of public land-use and disposition policies, see Appendix 7.

Urban Agriculture Infrastructure–Related Policies and Programs

Abandoned-Property Management Programs. Local and regional government agencies, together with nongovernmental organizations such as community development corporations (CDCs), neighborhood councils, environmental groups, and food security–related groups, can inventory, acquire, and dispose of vacant property for urban agricultural use. In order to implement efficient, effective, and sustainable programs, many communities have created abandoned-property management systems (APMS). APMS is "an organized process linking all of the activities involved in dealing with abandoned properties, from acquisition through disposition, in a way that maximizes the use of public resources, minimizes the present harm caused by those properties to the community and maximizes the city's redevelopment opportunities for the future" (Mallach 2006). This system paves the way for the inventory and acquisition of vacant property—the first step to providing land for urban agricultural use—and ensures the maximization of public and private resources and efforts.

Brownfield Cleanup Programs

Local governments exhibit an understandable reluctance to allow urban agriculture where they believe environmental hazards exist or where they might have to assume liability for environmental contamination. Despite these risks, many local governments are actively collaborating with community-based organizations and community development corporations to assess, remediate, and reuse brownfield sites for urban agriculture activities. The City of Lawrence, Massachusetts, partnered with Groundwork Lawrence, a nonprofit organization that works to engage residents in bettering local environmental conditions through community-building projects, to implement a citywide assessment of brownfield sites intended for community

Figure 3.11. Increasingly, brownfield sites like this one in Virginia are being reclaimed for urban agriculture uses.

Kimberley Hodgson

garden use. With funding from the U.S. Environmental Protection Agency (EPA), Lawrence conducted Phase 1 site assessments for more than 20 sites in underserved, low-income neighborhoods across the city. Groundwork Lawrence also worked closely with the EPA, the National Park Service, and landscape architects to design raised beds and other measures to minimize exposure for residents wishing to garden on sites not included in the citywide assessment. (See www.groundworklawrence.org.)

Local Procurement Policies

Because widespread urban agriculture is dependent on both market demand and a distribution and retailing infrastructure for local food, policies that increase grower-to-consumer linkages are another key part of a municipal urban-agriculture implementation strategy. These policies include local procurement commitments. One example is San Francisco mayor Gavin Newsom's 2009 executive directive instructing all city departments and agencies to purchase local and sustainably certified foods whenever possible (City and County of San Francisco Office of the Mayor 2009).

Local governments can pair these kinds of procurement policies with regulations pertaining to produce sold at farmers markets. For example, standards for the city-run farmers markets in Sacramento, California, stipulate that all produce offered for sale must have been grown within 25 miles of the city (Sacramento 2010). The largest farmers markets in Madison and Milwaukee, Wisconsin, are producer-only markets, ensuring that the food sold there is grown in state.

USING URBAN AGRICULTURE TO INFLUENCE THE OUTCOMES OF PRIVATE DEVELOPMENT PROJECTS

While city governments can advance urban agriculture through planning and policy, they also control mechanisms to review and guide private development in the public interest. Urban agriculture has yet to be widely incorporated into staff reviews of development projects, something that could result in negotiations with developers for more food-production sites. But as planners come to better understand its dimensions and benefits, urban agriculture may consistently be a part of future private developments as planning staff work with developers to ensure compliance with particular quality-of-life goals specified in local comprehensive plans.

Site Design and Development

Design Guidelines. For decades, local governments have been using design guidelines to ensure that new development projects enhance community character and are built in compliance with the goals and policies stated in the local comprehensive plan. During the development review process, planners and public officials check specific proposals against adopted guidelines to see how well these proposals conform to the community's vision for physical development. Traditionally, design guidelines have addressed building orientation and form, site design, signage, open space, and landscaping. They give local governments an opportunity to communicate expectations to developers and give developers a measure of certainty about what projects will or will not be approved.

The modification of existing design guidelines, particularly those applying to open spaces and landscaping, could encourage developers to incorporate food production into new projects. For example, in Minneapolis, developers who submit PUD proposals for review are given bonus points for incorporating green roofs and other growing spaces such as urban agriculture or community gardens into their designs (Minneapolis 2010). In January 2009, the Vancouver City Council adopted new urban-agriculture design guidelines as a key food-system policy for the city. The city's urban agriculture steering committee and green-building strategy technical team worked with landscape development staff to draft urban agriculture design guidelines for the private realm that would contribute to the city's larger green building strategy. Draft guidelines were reviewed by developers, landscape architects, urban gardeners, and the Vancouver Food Policy Council and subsequently presented to the public and then to the City Council.

The resulting document, *Urban Agriculture Guidelines for the Private Realm,* now provides guidance to urban designers, architects, landscape architects, planners, civic and environmental engineers, and private developers on the design and placement of urban agriculture and associated infrastructure in new private, primarily residential developments (Vancouver 2009). The guidelines include recommendations for the design and siting of shared garden plots and edible landscaping for public and private areas including patio, balcony, and roof deck spaces, as well as a range of supporting facilities such as storage space, composting facilities, and greenhouses. They also stress the importance of locating garden plots

Planned residential developments have increasingly incorporated community gardens, orchards, and urban farms as public amenities.

John Reinhardt

with other amenities such as covered outdoor shelters, children's play areas, community kitchens, and outdoor seating areas, to facilitate and encourage social interaction. Creating a separate set of guidelines was considered a more efficient approach to establishing urban agriculture policy in Vancouver than amending existing plan language. However, due to competition from other public benefit options, developers are not yet choosing to include shared garden plots or edible landscaping in new development proposals.

Development Projects. Adopting some of the concepts behind Ebenezer Howard's Garden City and Frank Lloyd Wright's Broadacre City models, planned residential developments have increasingly incorporated com-

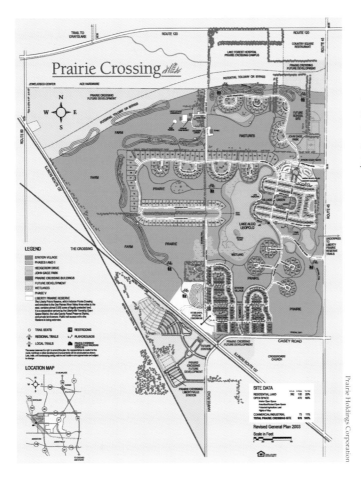

Figure 3.12. Planned residential developments like Prairie Crossing in Greyslake, Illinois, have incorporated agricultural features.

munity gardens, orchards, and urban farms as public amenities. Examples include the pioneering Village Homes of the mid-1970s in Davis, California (where 20 percent of the project's 70 acres contained row crops, vineyards, and orchards), or the more recent Prairie Crossing (Figure 3.12) in Grayslake, Illinois, 40 miles northwest of Chicago, and South Village in South Burlington, Vermont. Following these leads, a handful of communities have taken steps to encourage the integration of urban agriculture into future large-scale projects by adding uses such as community gardens and demonstration farms to lists of appropriate recreation or open-space features. Building on this idea, the latest version of the SmartCode model form-based code includes a module showing which types of urban agriculture fit in each transect (DPZ 2008).

AGRICULTURAL URBANISM

Agricultural Urbanism (AU) is a newly developing planning framework—promoted in different forms and under different labels by practicing architects, landscape architects, and planners—that sees municipal food networks as analogous to other vital infrastructure such as roads or sewers. It aims to improve food access, security, and knowledge by integrating context-sensitive urban agriculture and other food-related activities into a wide range of development settings (de la Salle and Holland 2010).

In one form, AU employs the new urbanist concept of the rural-to-urban transect to illustrate how different types of food production can be included not only in low-density residential areas, such as Prairie Crossing, but also in denser, urban neighborhoods (HB Lanarc n.d.). In another form, the European concept of Continuous Productive Urban Landscapes (CPULs) envisions planned combinations of connected urban open spaces combining urban agriculture with ecologically productive landscapes (Viljoen 2005). CPULs are characterized by their citywide scale and their aim to bring urban residents into close contact with social and natural activities and processes associated with nonurban locations.

AU proponents hope the framework will eventually inform comprehensive planning efforts, but it can also be used as a tool to help improve the design of new large-scale projects. For example, Vancouver incorporated AU principles into its development vision for the Southeast False Creek neighborhood. In 2007, the city released a study that outlined specific strategies for integrating urban agriculture and related support and management systems into a variety of public and private spaces in the Southeast False Creek neighborhood. The study is intended to provide both inspiration and guidance to project design and review teams as they work to implement the official development plan for the area (Vancouver n.d.).

SUPPORTING URBAN AGRICULTURE THROUGH PUBLIC-SECTOR PROGRAMS

Beyond supportive zoning regulations and other policies influencing public and private land use, a number of programs administered by local governments can be used to build the capacity of local growers or strengthen the infrastructure necessary for widespread, sustainable urban food production. These initiatives include community-garden programs, demonstration farms, municipal composting, education and technical assistance for growers, job training, grants, and direct-sale programs.

Municipal Community-Garden Programs

Municipal community-garden programs connect prospective gardeners and gardening groups with public land set aside for food or horticultural production. These programs also typically establish gardening standards and operating rules for participation. Some programs even provide supplies or technical assistance. Two of the oldest and most successful of these programs are found in Seattle and New York City. Seattle's P-Patch program, housed in the city's Department of Neighborhoods, currently oversees 73 gardens covering 23

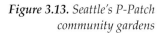

Figure 3.13. Seattle's P-Patch
community gardens

acres throughout the city (Figure 3.13). The program has been so popular over three decades that in 2008 residents approved a special levy of $2 million to develop new gardens. (See www.cityofseattle.net/neighborhoods/ppatch.) In New York, the Department of Parks and Recreation houses GreenThumb, the nation's oldest municipal community-garden support program. GreenThumb began in 1978 when community residents started gardening on sites abandoned during that decade's municipal fiscal crisis. It became a part of city government in 1995 and now supports more than 600 gardens across all five boroughs. (See www.greenthumbnyc.org.) Among its services, GreenThumb offers information on existing gardens through the city's Open Accessible Space Information System, a GIS application.

Publicly Owned Demonstration Farms

Publicly owned demonstration farms have an educational mission. Often, cities use these farms to teach residents about where their food is grown and to educate potential growers about sustainable farming practices. The farms may be operated by city staff or through a partnership with a nonprofit organization. For example, Zenger Farm in southeast Portland, Oregon, is a six-acre organic demonstration farm on land owned by the city's Bureau of Environmental Science (BES). In 1999, the BES entered into a 50-year lease agreement with the nonprofit Friends of Zenger Farm, which maintains the farm and runs educational programs on it. (See www.zengerfarm.org.)

Many local governments and universities also offer technical assistance or educational opportunities for growers. These programs range from informational brochures to multipart classes on sustainable gardening practices. Portland's Bureau of Planning and Sustainability recently began offering classes in gardening, urban chicken and beekeeping, cheese making, and canning (Portland Bureau of Planning and Sustainability 2010).

Municipal Composting Programs

A number of local governments have created municipal composting programs to separate food, yard, and garden waste from recyclable and nonrecyclable solid waste. These programs reduce the amount of organic waste being sent to landfills and provide nutrient-rich soil for gardeners. In 2004, San Francisco launched a mandatory citywide curbside composting program. (See Figure 3.14.) The city's green carts accept food scraps, food-soiled paper, and yard waste, diverting approximately 105,000 tons of refuse from landfills each year. After collection and processing, the compost is sold at area gardening-supply stores (City and County of San Francisco Department of the Environment n.d.). Large-scale composting as an entrepreneurial endeavor is discussed in Chapter 2.

Job-Training Programs

A renewed national interest in local and sustainable food has led to the creation of a number of job-training programs to teach city residents how to grow and sell food. Although most urban agriculture job-training programs are administered by nonprofit organizations, many receive public funding or operate on public land. One such program is Chicago's Growing Home (www.growinghomeinc.org), which offers such training to homeless and low-income residents on former federal land made available through the 1987 McKinney-Vento Homeless Assistance Act. In 2009, the City of Chicago announced it would be providing funding for 10 new jobs at Growing Home through the city's Community Green Jobs

Figure 3.14. San Francisco curbside compost-collection program flyer

City of San Francisco

program (Chicago Department of the Environment n.d). In Milwaukee, Growing Power has partnered with the Milwaukee Area Workforce Investment Board (MAWIB) to provide job-training opportunities for inner-city youth. For example, MAWIB trainees have assisted Growing Power staff in building new food-production hoop houses across Milwaukee as vehicles for developing construction skills appropriate for the emerging green economy.

Community Education Programs

Public and private health professionals, food-security organizations, and other community-based nonprofit organizations can play important roles in developing and implementing a variety of nutrition, health, food-literacy, and environmental-stewardship programs (Pothukuchi and Kaufman 2000; McCann 2006). The Ohio State University Extension Service for example, offers a variety of educational programs to urban agriculture practitioners and the general public. One of these, the Market Gardening Training Program, provides hands-on education in agricultural, business, and marketing skills required for a new business (Ohio State University Extension n.d). Denver Urban Gardens, a nonprofit organization, partners with a number of local organizations, including Denver Public Schools, Slow Food Denver, and Learning Landscapes, an open-space design program of the University of Colorado, Denver, to provide school classes in biology, ecology, horticulture, wellness and nutrition, recycling, composting, and community building. With Denver Recycles, a program of Denver Public Works/Solid Waste Management, it also offers free public composting classes. (See http://dug.org/education.)

Municipal Grant Programs and Other Financial Assistance

Some local governments have also provided grants for other types of urban agriculture activities. For example, Toronto's Environment Office funds urban agriculture projects through two grant programs: the Live Green Toronto Community Investment Program (www.toronto.ca/livegreen/greenneighbourhood_rebates_cip.htm) and the Community Service Partnerships program (www.toronto.ca/grants/csp/whats-new.htm). The first provides funding to grassroots initiatives to reduce greenhouse gas emissions, and the second provides funding for gardens, markets, and kitchens in neighborhoods where access to fresh local food is poor. In order to offset the start-up costs for entrepreneurial commercial urban agriculture within Cleveland, the Cleveland Department of Economic Development initiated "Gardening for Greenbacks." This program provides up to $3,000 in grant funding to people who have completed the Market Gardening Training Program; these grants can be used for tools, display tables and booths, irrigation systems, rain barrels, greenhouses, and signage.

Some cities have used a portion of their discretionary federal Community Development Block Grants (CDBG) for urban agriculture. A small percentage of Madison's CDBG funding is dedicated to a New Garden Fund, administered in collaboration with the Community Action Coalition of South Central Wisconsin. Under this program, a community group can receive up to $2,000 to start a new garden, expand an existing one, or relocate one threatened by development. And since 1985, Boston's Grassroots Program has provided CDBG funds for technical assistance to and capital construction on new and existing garden sites.

Direct-Sale Programs

A variety of direct-sale models provide urban food growers with the opportunity to sell their products directly to local businesses, institutions, and consumers. Though local governments are not directly involved in these private market transactions, they can still play a role in encouraging some of these programs through policies that support farmers markets, on-site markets, and farm-to-institution programs.

Farmers Markets and On-Site Markets. Farmers, artists, and other vendors meet at farmers markets to sell their products directly to consumers (Figure 3.15). Such markets provide access to local food and can serve as community gathering spaces, hosting activities such as demonstration cooking and gardening workshops and live music performances. They connect local producers with consumers and can build social capital in neighborhoods (Groc 2008).

Figure 3.15. *The Renton, Washington, farmers market brings farmers and consumers together in a community gathering space.*

Public Health–Seattle/King County

The recent rapid increase in the number of urban farmers markets across the United States and Canada (USDA AMS 2010) has created significant direct-marketing opportunities for urban and periurban farmers, yet relatively few local zoning codes acknowledge farmers markets as a permitted land use. Consequently, these markets often operate either without formal sanction or through a temporary use permit. In both cases, even the most successful markets could be displaced if a more profitable land use is proposed. A growing number of communities have revised zoning standards to formally acknowledge farmers markets as permitted uses in certain districts, and model language to zone markets is available for communities to use (NPLAN 2009). Other communities have used special administrative provisions to sanction the use of city-owned property for farmers markets. In Philadelphia, public or private entities may operate farmers markets on public rights-of-way, subject to the standards and licensing procedures outlined in the city code (Philadelphia 2010a). For additional policy examples, see Appendixes 5 and 7.

The on-site market, at which growers sell their products at or near the location of their garden or farm plot, is another direct-sale method used by urban growers. The Troy Gardens Community Farm in Madison regularly sells its seasonal produce from a weekly farm stand set up on a bordering avenue. However, land-use regulations often prohibit on-site sales, particularly in residential zoning districts. The City of Cleveland recently updated its zoning code to allow on-site sales from community and market gardens as a permitted main use. (See Appendix 5 for more information.)

To reach areas with little or no access to grocery stores or fresh fruit and vegetable markets, some urban and rural farmers deliver produce directly to consumers via refrigerated trucks. In upstate New York, the Capital District Community Gardens' Veggie Mobile operates like an ice-cream truck, stocked with a variety of fresh fruits and vegetables. The Veggie Mobile

Local markets allow urban growers to expand their channels of distribution beyond traditional outlets, generating a stable source of farm-based income and improving the local economy by increasing the economic viability of the farm sector.

visits senior and assisted-living centers, public housing projects, and other densely populated locations in Albany, Schenectady, and Troy. It is equipped with refrigerators and shelves, rooftop solar panels, a biodiesel engine, and a sound system to announce its arrival. (See www.cdcg.org/VeggieMobile .html.) Since mobile vending is also regulated by local governments, planners can play a role in revising existing or establishing new policies to allow the mobile vending of locally grown fresh fruit and vegetables.

Farm-to-Institution Programs. A farm-to-institution program abets the direct sale of locally produced food products to schools, universities, and colleges, hospitals and long-term care facilities, prisons and correctional facilities, and other institutions (Bellows, Dufour, et al. 2003). These local markets allow urban growers to expand their channels of distribution beyond traditional outlets, generating a stable source of farm-based income and improving the local economy by increasing the economic viability of the farm sector. They can also create opportunities for added income by offering educational experiences to clients; help improve the nutritional quality of institutional meals; and re-create relationships in the community between consumers and farmers. Because institutional dining facilities already use many fruits and vegetables produced by small farmers, these programs offer an opportunity to create a permanent and increasing demand for locally produced food products.

Kimberley Hodgson

Farm-to-school programs in particular offer multiple benefits for farmers, students, and schools. They not only connect small farmers with area schools but also aim to (1) enhance the quality and quantity of fruits and vegetables available to students; (2) increase students' understanding of, connection to, and appreciation of food and its progress from farm to plate; (3) provide a context for nutrition education; and (4) increase the economic viability of the local farm sector (Tropp and Olowolayemo 2000; Harmon 2003). Farm-to-school programs are best served by public policies that provide uniform objectives and standardize practices across a number of schools. Such policies, such as the pioneering School Lunch Initiative of the Berkeley Unified School District (2004), support strong roles for food-producing schoolyard gardens within a larger set of curricular actions (Rauzon et al. 2010). Other policies, such as the 2010 partnership between Growing Power and Milwaukee Public Schools, aim to direct the produce of urban and periurban farms directly into school cafeterias.

CONCLUSION

Traditional planning practice—the techniques and approaches understood by most planners—can be used for the purpose of advancing urban agriculture within local jurisdictions. Yet while processes, policies, programs, and initiatives that support or provide incentives for urban agriculture through planning practice are developing, we do not yet have a complete picture of the full potential of a marriage of urban agriculture and planning. Chapter 4 reflects on the connections between urban agriculture and the desire of planners to take on big issues, such as urban redevelopment, sustainable growth, and resilience, as well as greater food security in cities.

CHAPTER 4

Linking Urban Agriculture
with Planning Practice

 Planners are increasingly recognizing urban agriculture as an important component of sustainable and resilient environments. The 21st century presents a range of challenges both new and perennial, including sustainability, disaster recovery, climate mitigation and adaption, urban revitalization, and economic development. Local food production can be an important complement to planning strategies that address community building, environmental health, food security, stormwater management, and jobs generation. Through case-study research, this chapter links urban agriculture with many different areas of planning practice.

APA conducted extensive case-study research on how U.S. and Canadian cities are planning for urban agriculture and connecting it to broader areas of planning practice.

Between March and September 2010, APA conducted extensive case-study research on how U.S. and Canadian cities are planning for urban agriculture and connecting it to broader areas of planning practice. APA interviewed urban agriculture practitioners and advocates, local government officials, and planners in 11 cities in North America: Chicago, Cleveland, Detroit, Kansas City (Kansas and Missouri), Milwaukee, Minneapolis, New Orleans, Philadelphia, Seattle/King County (Washington), Toronto, and Vancouver.

As part of this research, APA asked participants about the background and history of their metropolitan regions, the major urban-agriculture stakeholders and actors, and the extent of collaboration among them. Additional questions addressed local government and planning contexts (e.g., whether comprehensive planning was state mandated), whether any community food assessments or other food-related studies had been done in the jurisdiction, and what local policies and programs might have impacts on urban agriculture. APA also asked questions about brownfields assessment and remediation for urban agriculture uses.

Nevin Cohen

Selected cities featured strong urban-agriculture practitioner communities (e.g., Milwaukee, Chicago, Philadelphia), long histories of community food-systems work (e.g., Toronto, Vancouver, Seattle/King County), or recent innovations in processes, plans, or land-use regulations (e.g., Minneapolis, Cleveland, Kansas City). New Orleans was chosen for its need to rebuild its food system. Some cities are nearly built out (e.g., Minneapolis, Vancouver, Seattle/King County), while others are in the process of redeveloping and renewing their centers (e.g., Cleveland, Detroit, Philadelphia, New Orleans).

FOSTERING RESILIENT COMMUNITIES

> In a resilient city every step of development and redevelopment of the city will make it more sustainable: it will reduce its ecological footprint (consumption of land, water, materials, and energy, especially the oil so critical to their economies, and the output of waste and emissions) while simultaneously improving its quality of life (environment, health, housing, employment, community) so that it can better fit within the capacities of local, regional, and global ecosystems. Resilience needs to be applied to all the natural resources on which cities rely.
>
> In resilience thinking the more sustainable a city the more it will be able to cope with reductions in the resources that are used to make the city work. Sustainability recognizes there are limits in the local, regional, and global systems within which cities fit, and that when those limits are breached the city can rapidly decline. The more a city can minimize its dependence on resources such as fossil fuels in a period when there are global constraints on supply and global demand is increasing, the more resilient it will be. (Newman, Beatley, and Boyer 2009)

Resilient cities are able to adapt to changes that stress their social, environmental, and economic systems; they create redundancies and alternative systems to respond to challenges such as peak oil, climate change–induced weather patterns, and economic downturn. While there is a strong foundation for sustainability in federal environmental law and the recent federal Partnership for Sustainable Communities among the Environmental Protection Agency, the Department of Transportation, and the Department of Housing and Urban Development, local governments will likely need to continue taking responsibility for enhancing their own resilience (Daniels 2008). As the following case studies demonstrate, U.S. cities use urban agriculture to varying degrees in fostering resilience.

The vertically integrated, heavily consolidated, industrialized food system contributes to a lack of resilience in both urban and rural communities (Pothukuchi and Kaufman 1999; Roberts 2008).[1] Urban agriculture can be one tool in increasing local flexibility in responding to crises.

Numerous cities are incorporating urban agriculture in municipal planning and policy making to increase their long-term sustainability and resilience. (See also Chapter 3.) The approaches taken are driven by political leadership and vision, grassroots advocacy from the urban agriculture community, and boundary-spanning and bridging work from the nonprofit sector (Caton Campbell 2004; Stevenson et al. 2007). Charismatic leaders such as 2008 MacArthur Foundation Fellow and urban farmer Will Allen, founder and CEO of Growing Power (www .growingpower.org) in Milwaukee, create fertile ground for extreme innovation in the practice of urban agriculture, with city plans and policies catching up later. The following section explores the integration of urban agriculture into citywide sustainability planning efforts in Philadelphia, Toronto, and Vancouver.

Philadelphia has a long-standing tradition of gardening on vacant or underutilized lands to stabilize neighborhoods and foster a sense of community.

Philadelphia

Despite being a major center of urban agriculture, Philadelphia (pop. 1,446,395) is among many cities where local government is attempting to "catch up" with the nonprofit and informal community gardening and farm-

ing activity in its neighborhoods. Philadelphia has a long-standing tradition of gardening on vacant or underutilized lands to stabilize neighborhoods and foster a sense of community. Two very robust urban-agriculture support systems have existed in the city for several decades: the Penn State Urban Gardens Program and the Pennsylvania Horticultural Society (PHS) Philadelphia Green Program, which includes the City Harvest and City Harvest

Though Philadelphia's community-gardening support systems were defunded and gardening declined in the 2000s, some city bureaucrats took important steps to enable urban farming.

Growers Alliance, sometimes complemented by the Neighborhood Gardens Association land trust. These programs link urban farmers with a variety of distribution outlets. And in recent years, nationally recognized farming and food projects—such as Greensgrow Farm, Mill Creek Farm, Weaver's Way, and the Philadelphia Orchard Project—have earned the city a reputation as a hotbed of urban agriculture activity.

Nevertheless, prior to 2008 the city lacked anything resembling a comprehensive vision for food and agriculture policy. In the 1980s and 1990s, various city departments, including the Office of Housing and Community Development and the Recreation Department, intermittently supported PHS community gardening programs. The Redevelopment Authority (RDA) managed an adopt-a-lot program that was also uneven in its oversight. The city government did not establish a public-sector garden-support program, nor did it formalize community gardening as a land use, either in the zoning code or through land preservation.

In the early and mid-2000s, though the city's community-gardening support systems were defunded and gardening declined, some city bureaucrats took important steps to enable urban farming. The Neighborhood Transformation Initiative, the urban renewal program of Mayor John Street, bulldozed many smaller gardens as it assembled land intended for development. But the Philadelphia Water Department's economic development director, Nancy Weissman, initiated a pilot project that helped create the Somerton Tanks Farm, a three-year experiment that demonstrated the Small Plot Intensive farming method. The department's Office of Watersheds tested the viability of urban agriculture for stormwater management by supporting the establishment of Mill Creek Farm in West Philadelphia, a small farm dedicated to community education and food access. Beginning in 2007, Joan Blaustein, director of land management for the city's park system, set about assessing park properties' suitability for urban agriculture. She subsequently advanced a vision of urban farmers gaining access not only to vacant lots but also to the city's best farmland, including three historic farms, each over 50 acres, in city parks.

In 2008, new mayor Michael Nutter established the Mayor's Office of Sustainability and charged its first director, Mark Alan Hughes, with developing a comprehensive sustainability plan. That summer, a group of nonprofit and academic leaders lobbied Hughes to include local food and agriculture policy in his office's initiatives. In October, the office issued a food charter by executive order, becoming the first municipal government of a large U.S. city to do so. This charter asserted community food security as a basic right of all Philadelphians. It encouraged the expansion of local food production, processing, distribution, and waste management. The Office of Sustainability and the park system soon purchased a large digester to revamp the city's public composting operations. The charter also outlined the city's aims to create a municipal food-policy council, revise zoning and vending codes to support urban agriculture, and open public land to farming and gardening.

The following spring, the Office of Sustainability unveiled Greenworks Philadelphia, a sustainability plan that had the broad goal of making the city the greenest in America. One of the sustainability targets in Greenworks is ensuring that 75 percent of city residents have local food within a 10-minute walk by 2015. Under this target are subgoals to start 12 commercial farms, 15 new farmers markets, and 59 new food-producing gardens in the city. At this writing, the PHS City Harvest Growers Alliance project and urban farmers have achieved the first goal. The city's farmers markets are expanding, though apart from the new market at city hall they are mostly managed by the nonprofit Food Trust, not by city government.

The city has released a draft of its new zoning code, which defines and recognizes community gardens, community-supported agriculture (CSA), market farms, and animal husbandry as primary or accessory uses in a broad variety of zoning districts. However, it prohibits farms and community gardens in low-density residential districts (R-1), and the city's new draft lease agreement for community gardens on public land requires gardeners to carry insurance at the level of built-on property. Urban agriculture practitioners have criticized these proposals as overly restrictive and burdensome. As the Planning Commission embarks on producing its first comprehensive plan in 50 years, planners will have an opportunity to articulate a long-term vision for urban agriculture and food access. However, Philadelphia faces significant challenges in its shifting landscape of community gardens and farms.

The Greenworks plan made clear that various city departments, including planning, parks and recreation, health, water, and the RDA, have important roles to play in fostering and regulating urban agriculture. But these departments' divergent missions and visions for urban agriculture raise two important questions. First, what forms should urban agriculture take, particularly as a land use? Second, which departments should take the lead in defining the city's vision and approach to supporting urban agriculture? The answers to these questions are still being worked out in Philadelphia.

Perhaps the most fundamental disagreement over the form of urban agriculture concerns whether farming and gardening should be approached as long- or short-term land uses in the city. The RDA views urban agriculture as an interim use, consistent with its mission to dispose of land to increase the local tax base and the supply of affordable housing. Its director, Teresa Gillen, was attracted to urban agriculture as a way of stabilizing vacant land and making it more attractive to developers. In 2009, the agency issued a request for proposals for urban farmers to grow on RDA land for a maximum of three to five years. After receiving tepid response from farmers, most of whom saw little reason to invest in sites and soils only to be displaced in a few years, the agency retracted the RFP. In 2010, the RDA, Planning Commission, and Office of Sustainability pursued further studies of how they might manage community gardens and urban farms as interim uses. The RDA and Philadelphia Housing Authority, however, failed in an attempt to build new housing on Mill Creek Farm and the adjacent Brown Street Garden, where several decades earlier houses had fallen into the collapsed sewer that carries the historic creek directly below the site. After two years of local advocacy to save the farm, the U.S. Department of Housing and Urban Development determined it would not support the housing project because the site hosts urban agriculture.

Philadelphia's Department of Parks and Recreation and grassroots advocates have advanced alternative visions of urban agriculture as a more stable, long-term land use that promotes health, recreation, and environmental sustainability rather than serving short-term economic and redevelopment goals. In 2009, the department supported the development of the city's largest local CSA farm operated by the for-profit Weavers Way Cooperative at Saul Agricultural High School, where the farming operations of the school are on parkland. Worked by Weavers Way farmers, Saul students, and CSA members, the CSA is integrating sustainable, chemical-free agricultural education into what is mainly a technical school for industrial agriculture.

However, the Department of Parks and Recreation's broader plans for transitioning the city's large farms from industrial pro-

In 2009, Philadelphia's Department of Parks and Recreation supported the development of the CSA farm operated by the for-profit Weavers Way Cooperative at Saul Agricultural High School, where the farming operations of the school are on parkland.

Nevin Cohen

duction of hay, feed corn, and livestock to chemical-free, more labor-intensive farming of food for people suffered a setback in 2010. A small group of affluent neighbors of Manatawna Farm, a 76-acre property that includes hay farming and pastureland for Saul as well as a five-acre community garden, opposed the department's plan to open five acres to 10 urban farmers seeking land for sustainable food and flower production. The neighbors and representatives of Saul convinced the city council to pass a bill outlawing commercial farming at Manatawna, based mainly on the argument that the hay fields are important bird habitat. (Affluent communities in the past have used similiar conservation claims to prevent development.) Weeks later, the council passed a virtually identical bill preventing commercial farming on the Philadelphia side of the park's largest agricultural property, 112-acre Fox Chase Farm, which straddles the city line.

Community gardening in Philadelphia faces considerable challenges and open questions about its future. Though gardens remain by far the largest part of urban agriculture in Philadelphia, the city government has no strategy for their preservation or management. For decades, the government has relied on the PHS to support gardens, but PHS community-garden programs have shrunk. In 2010, the outgoing leadership at PHS decided to close the Neighborhood Gardens Association land trust, though the new president reversed that decision. The Office of Sustainability took a modest step in 2010 when it expanded its website with pages directing people interested in gardening to PHS programs and other nonprofit resources. However, Philadelphia lacks a citywide system for enabling residents to locate plots in community gardens, as cities such as Seattle do. Also in 2010, the city's Department of Aviation declined to renew the lease for the city's largest community garden by far, the 11-acre Airport Garden, because of potential airport-expansion issues.

In summary, Philadelphia presents some of the most vital and promising nonprofit and public-sector initiatives in urban agriculture but also some of the greatest challenges. The city's food charter and sustainability plan articulate a strong vision, yet it remains unclear which city departments should determine specific urban agriculture policies and lead their implementation. While regulatory barriers are being removed, planners and their colleagues in city government have yet to tackle important questions of land tenure and public access. In November 2010, the Philadelphia Department of Public Health hired a food policy coordinator, bringing one more department into the mix. This position is funded by a two-year grant from the Centers for Disease Control, raising another important question for planners in Philadelphia and other cities; namely, how city governments can institutionalize their capacity to support urban agriculture beyond the current wave of popular attention and funding for it.

Toronto

In Toronto, Ontario (pop. 2,503,280), the Environment Office is the lead agency behind the city's efforts to address climate change and greenhouse gas emissions. The 2007 Climate Change, Clean Air and Sustainable Energy Action Plan linked local food production and urban agriculture (such as community gardens, local food markets, and the city's food purchasing policies) to a host of other climate-change efforts (such as reducing energy consumption), specifically through the creation of an interagency working group convened by the Environment Office with representatives of the Board of Health, Toronto Community Housing, the Economic Development Division, and the Toronto and Region Conservation Agency. (See City of Toronto 2007.) The plan also called for the creation of an Enviro-Food Working Group to promote local food production, remove barriers to local

Toronto's Environment Office is the lead agency behind the city's efforts to address climate change and greenhouse gas emissions.

markets, and review city food-purchasing policies. The group is working to identify the barriers to urban agriculture in Toronto and will report its findings to the City Council.

The Environment Office is the home of Live Green Toronto, a program that offers funding and incentives to neighborhood-based greening projects through Community Investment Program grants (for which a community group may receive up to C\$25,000). Although urban food production is just one type of green activity that can be supported by such grants, seven of the 13 awards in 2009 were for urban agriculture projects. (See www.toronto .ca/livegreen/greenneighbourhood_rebates_awards.htm#cig2009spr.) With additional funding from other city agencies, the Environment Office also sponsors the Food Security Investment Program centering on Community Food Animators, or leaders from nonprofit food advocacy groups, including representatives from FoodShare (www.foodshare.net), The Stop Community Food Centre (www.thestop.org), and the Afri-Can Food Basket (www.afri-canfoodbasket.com), who perform urban agriculture outreach focused on training assistance, funding opportunities, and regulatory issues.[2]

Vancouver

In Vancouver, British Columbia (pop. 578,000), urban agriculture is an important part of broad sustainability and food-systems agendas. The city's rich history of food-related programs, services, and planning activities has created a foundation for over a decade's worth of policy innovation and reform. Vancouver's strengths are rooted in its comprehensive approach to the integration of food-system considerations in municipal policy decision-making processes and its understanding and explicit acknowledgment of the connections between urban agriculture and sustainability, neighborhood livability, urban greening, community building, social interaction, and crime reduction.

As a result of decades of research and advocacy work completed by a variety of food-related organizations, on July 8, 2003, the Vancouver city council approved a motion "supporting the development of a just and sustainable food system for the City of Vancouver." The mandate prompted the council to establish a food policy task force comprised of two councilors, one school board trustee, one board member of the parks and recreation commission, and representatives from Vancouver Coastal Health, the Greater Vancouver Regional District, and more than 70 community groups. (See Forum of Research Connections et al. n.d.)

Between July and December 2003, the task force developed the Vancouver Food Action Plan, which established three main goals for how the city should satisfy the council's mandate: (1) create a food policy council; (2) develop an interim work plan; and (3) develop an implementation support system. The Vancouver city council adopted the action plan in December 2003. To achieve the plan's goals, the council established the municipally affiliated Vancouver Food Policy Council (VFPC) to serve as an advisory group to the council and function as a bridge between citizens and civic officials; initiated a comprehensive assessment of Vancouver's food system assets and gaps; and created two new food-policy staff positions—the first food-system planning positions in a North American local government.

These positions included one full-time position (food policy coordinator) and one temporary (two-year) full-time position (food system planner). The purpose of the coordinator was to act as the liaison between the VFPC and city government, while the planner was to "internally coordinate and implement existing and new food-related programs and services" within Vancouver (Mendes 2008). A few months after the adoption of the plan, the city council approved the plan's expenditures. On July 14, 2004, the food policy task force stepped down and handed over the food policy responsibilities to the newly elected members of the VFPC.

Vancouver adopted hobby beekeeping good-management practices in 2006 to support hobby beekeeping within city limits and ensure that it is a safe and suitable activity for residential areas.

These initiatives increased the City of Vancouver's "institutional capacity to implement food policy" and provided a solid foundation for subsequent political, institutional, and public support of urban agriculture planning, goals, standards, policies, and projects (Mendes 2008). Since the mandate, the city has made great progress in improving the regional food system, including its urban food production. Table 4.1 provides an outline of the city's major accomplishments between the adoption of the citywide food-system mandate in 2003 and current legislation recently adopted in 2010.

Several events and policies mentioned in Table 4.1 significantly contributed to Vancouver's success in supporting a range of urban agriculture activities within city limits: the Food Action Plan; the development of the VFPC; the reform of the animal control bylaw to permit hobby beekeeping within the city; the 2,010 Garden Plots by 2010 initiative; the formation of an intergovernmental urban agriculture steering committee; the adoption of urban-agriculture design guidelines (see Chapter 3); and the development of new animal control and zoning and development bylaws to allow the keeping of backyard chickens (City of Vancouver 2005, 2009, Animal Control n.d., and Community Services 2010).

Vancouverconvention. Used under a Creative Commons license

Hobby Beekeeping. On February 27, 2006, the City of Vancouver adopted hobby beekeeping good-management practices to "support hobby beekeeping within city limits" and ensure that "hobby beekeeping is a safe and suitable activity for residential areas." It also eliminated the language in the animal control by-law that had prohibited hobby beekeeping. These municipal guidelines complement the British Columbia Provincial Bee Act. Together the provincial and municipal standards work to maximize the environmental benefits of hobby beekeeping, while minimizing the health risks.

Collaboration among city staff, regional environmental-health department staff, and the British Columbia Ministry of Agriculture was essential not only to prompting these policy changes but also to assuring the city council that public outreach and community education about the new guidelines would occur, and licensing, enforcement, and monitoring of complaints would be addressed (City of Vancouver 2005).

2,010 Garden Plots by 2010 Initiative. On May 30, 2006, city councilor Peter Ladner issued a challenge to individuals, families, community groups, and neighborhood organizations to establish more food-producing gardens in Vancouver. The motion presented before the council "for the City to work

TIMELINE	FOOD POLICY
July 8, 2003	City Council motion supporting the development of a just and sustainable food system for the City of VancouverWater hoses, rain barrels, and other equipment used to irrigate the garden or farm
December 9, 2003	City Council approved the Food Action Plan developed by the Food Policy Task Force
March 11, 2004	City council voted to approve the expenditures associated with the Action Plan
July 14, 2004	Food Policy Task Force, as its final act, elected members of Vancouver's first municipally affiliated food policy council
September 20, 2004	First meeting of the Vancouver Food Policy Council (VPFC)
December 2004–October 2005	Vancouver Food System Assessment conducted
July 25, 2005	VFPC and food policy city staff presented a report to City Council that requested an amendment to the health bylaw to allow for hobby beekeeping within the city
September 13, 2005	City council approve the removal of sewer-service charges for community gardens
September 19, 2005	Vancouver Board of Parks and Recreation adopts the Community Gardens Policy
October 27, 2005	Vancouver Food System Assessment released (see Chapter 3)
February 27, 2006	City council adopts Hobby Beekeeping Guidelines
May 30, 2006	Councilor Ladner announced the 2,010 Garden Plots by 2010 Initiative
January 2007	City Council adopts the Vancouver Food Charter
2007–2008	VFPC initiated a two-year study to "identify, review and analyze key factors that are required to support and enhance Vancouver's food security" and "identify key determinants of food security, denote benchmarks, and recommend strategic priorities and policies to be considered by the City."
2008	City staff establish the Urban Agriculture Steering Committee
January 20, 2009	City Council adopts Urban Agriculture Design Guidelines for the Private Realm
March 2009	City Council instructed staff to develop policy guidelines for the keeping of backyard chickens in Vancouver
June 10, 2010	City council amended the animal control bylaw and zoning and development bylaw to permit the keeping of chickens with the city

Sources: Mendes 2008; http://vancouver.ca/commsvcs/socialplanning/initiatives/foodpolicy/tools/pdf/councilmotion.pdf; City of Vancouver, Community Services 2009; Forum of Research Connections et al. 2005.

Table 4.1. Timeline of food policy in Vancouver, British Columbia

with the Vancouver Food Policy Council to encourage the creation of 2,010 new garden plots in the city by January 1, 2010, as an Olympic legacy" passed unanimously (City of Vancouver Community Services 2010).

In the motion, Ladner stated that "community gardens and other forms of urban agriculture are important neighborhood gathering places that promote sustainability, neighborhood livability, urban greening, community building, intergenerational activity, social interaction, crime reduction, exercise and food production." (See http://vancouver.ca/ctyclerk/cclerk/20060530/documents/motionb2.pdf.) The motion charged the city of Vancouver to collaborate with the VFPC, the school board, the Board of Parks and Recreation, community groups, neighborhood organizations, nonprofit groups, and individual citizens to create the new plots. Before the initiative, only 950 community garden plots existed throughout the city. Between 2006 and December 31, 2009, the city more than doubled that number: 1,079 new plots were added, bringing the total to 2,029. These gardens were developed on city-owned land, including parkland, but also on private property, with private developers incorporating community-shared urban agriculture spaces into the design of new residential developments.

Urban Agriculture Steering Committee. In response to the 2010 urban garden challenge, City Councilor Andrea Reimer and several city departments established an urban agriculture steering committee to ensure cross-departmental communication and increased coordination and collaboration in the development of new or revised policies, programs, and projects related to urban agriculture. This committee allowed individual departments to address specific issues related to urban agriculture, while understanding how the individual programs, projects, and policies would collectively support it. City staff "recognized that inter-departmental coordination [was] needed to address issues of urban agriculture … and integrate programs." The committee initially met quarterly but currently meets monthly to exchange information, share ideas, and resolve any issues. The committee is composed of senior city staff, including the director of social policy and the director of the Vancouver Park Board's East District, and the director of planning, the assistant director of the Development Services Inquiry Centre, the manager of sustainability, and several other environmental, social, and transportation planning staff. (See City of Vancouver, Vancouver Food Policy Council 2009.)

Chicken Keeping. Due to public pressure to legalize the keeping of backyard chickens within city limits, in March 2009 the city council instructed staff to develop policy guidelines for this use. On June 8, 2010, the council amended two important bylaws—animal control and zoning and development—to reverse the prohibition on backyard chickens. The amendments allow the keeping of hens and require residents to register each hen with the city and abide by certain rules and regulations related to the safe and humane keeping of hens in urban spaces. (See http://vancouver.ca/ctyclerk/cclerk/20100408/documents/penv3.pdf.)

Challenges and Lessons. Despite its success, Vancouver has also experienced its share of setbacks. Unlike many cities in the American Midwest that have lots of vacant land, Vancouver is a dense, built-out city, geographically confined by mountains and water. Expensive land and development pressures have created real barriers to the expansion of urban agriculture beyond community and private gardens.

While the majority of Vancouver's urban agriculture efforts have focused primarily on community gardening, edible landscaping, and other forms of noncommercial urban agriculture, the city has also begun to explore other forms of urban agriculture as public benefits in private development projects and as commercial enterprises. Due to popular press and an increased interest in urban agriculture, the interests of residents, city staff, and the city council have shifted considerably in the past year, explains Wendy Mendes, adjunct professor at the School of Community and Regional Planning at the University of British Columbia and former social planner and food systems planner for the City of Vancouver. "There is more of [an] explicit focus on urban farming. Not just community gardens, but social enterprise, community-supported agriculture, farming for profit, skill building, and employment training." Community gardens have existed in Vancouver for decades, but "until recently urban farming did not exist in Vancouver," explains Mendes.

While the concept, means, and scope of implementation of commercial agriculture within Vancouver city limits are still evolving, the city has provided grant funding for an experimental commercial urban farm. SOLEfood Farm (http://1sole.wordpress.com) is located on private property on Vancouver's Downtown East Side, one of the city's most disadvantaged neighborhoods. The farm, made up of hundreds of planters, offers training and employment to neighborhood residents. The food produced is sold to restaurants and at farmers markets and donated to community organizations focused on improving community food security. The city hopes to help support future urban farming enterprises by providing access to city-owned land and, potentially, additional funding. "What land do we have available? How can we increase land tenure? What are the different models of production? Conversations within municipal government are beginning to address these questions," says Mendes.

Vancouver has provided grant funding for an experimental commercial urban farm that is made up of hundreds of planters and offers training and employment to neighborhood residents.

SOLEfood

The events, policies, and programs mentioned above demonstrate the City of Vancouver's sustained commitment to achieving the 2003 Vancouver Food Action Plan's mandate supporting the development of a just and sustainable food system. (See City of Vancouver, Vancouver Food Policy Task Force 2003.) However, the city continues to struggle with finding the appropriate balance among regulation, formalization, and grassroots efforts. Mendes asks, "How much needs to be embedded within local government and how much should reside at the community level? [Determining] this remains a constant challenge."

Vancouver has also learned that comprehensive action is required to make real, lasting changes. The food system is complex and therefore necessitates an interdisciplinary, collaborative effort to simultaneously address multiple areas. "We can't solve these problems with planning alone. We need fewer silos and more cross-departmental collaboration, as is needed for all sustainability issues," explains Mendes. "We are lucky because of our strong history of food. We have also learned to listen well—involve and give a voice to our local experts and community activists. But we still have a lot more to do. If we really want to change the food system, we will need to look beyond policies and programs to changes to the built environment. The built environment is the big ticket item. At the end of the day we are trying to change the built form."

RECLAIMING VACANT LAND

The volume of vacant and abandoned property presents numerous opportunities for urban agriculture as a viable land-management strategy.

Aging or postindustrial cities, in particular, can have large amounts of vacant, underutilized, or contaminated sites that present multiple challenges for reuse. Sites range from small individual parcels to larger, aggregated tracts (Kaufman and Bailkey 2004; Vitiello 2008). Former uses also vary; in some cases, residential abandonment has left block upon block of dilapidated houses, while other cities suffer from industrial abandonment. Some cities, such as Detroit, Cleveland, and Chicago, have experienced both scenarios. Many former industrial or commercial sites are categorized as brownfields—sites that are contaminated or are perceived to have contamination from historic uses—which makes reuse especially challenging.

Though typically viewed as a problem, vacant land can become an important community asset if identified and rehabilitated. One important yet underutilized strategy is the reuse of land for open space or parkland, which focuses on productive land management and resource provision rather than redevelopment. However, urban agriculture is often overlooked as a reuse strategy. The volume of vacant and abandoned property presents numerous opportunities for urban agriculture as a viable land-management strategy. Areas plagued by abandonment, crime, trash, and weeds can be transformed into flourishing, colorful, and agriculturally productive open spaces that provide immediate economic, environmental, and health benefits. Abandoned buildings, if structurally sound, can be reused for food production, processing, distribution, or disposal purposes: as seed banks, community or commercial kitchens, food cooperatives, tool sheds, barns or other animal shelters, henhouses, farmworker housing, or in some cases, greenhouses.

Some cities are already making great strides in this direction, though different contexts present different challenges and opportunities for success, as illustrated by the cases of Detroit and Cleveland.

Detroit

In Detroit (pop. 971,121), vast acres of vacant land have created a desolate urban landscape. With approximately 50 of its total 138 square miles vacant, Detroit has become synonymous with urban disinvestment and postindustrial decline. While there is virtually no demand for new residential, commercial, or industrial land in the city, Detroit has gained a well-deserved national—and international—reputation as a haven for urban agriculture projects. In addition to hundreds of backyard gardens, the city is home to more than 600 community, school, and institutional gardens. And these tallies do not account for the guerrilla gardens and small farms where neighbors use land without obtaining permission from its owners.

Laura Buhl, AICP, a planner for the Detroit City Planning Commission, concedes that much of the city's urban agriculture has taken root without any official sanction. To date, Detroit has not adopted planning policies or land-development regulations that explicitly allow or encourage urban agricultural activities. Apart from listing commercial greenhouses as permitted uses in business and industrial districts, the city's code does not define or regulate community gardens, urban farms, or other related uses.

Since the 1970s, the city's primary tool for supporting urban agriculture has been the Farm-a-Lot program, which provides seeds and free tilling to residents who want to garden on city-owned lots next to their homes. Meanwhile, numerous educational institutions and nonprofits have responded to a demand for supplies and technical assistance. In 2004, four of these

Detroit's primary tool for supporting urban agriculture has been the Farm-a-Lot program, which provides seeds and free tilling to residents who want to garden on city-owned lots next to their homes.

organizations—the Greening of Detroit, the Detroit Agricultural Network, Michigan State University Extension, and Earthworks Urban Farm—used a USDA food security grant to fund the Garden Resource Program. By its own count, this partnership provides support to more than 875 urban gardens and farms in Detroit and two communities it encircles, Hamtramck and Highland Park. (See www.detroitagriculture.org.)

Farming in Detroit gained national attention in early 2009 when multimillionaire John Hantz announced plans to develop the world's largest urban farm on hundreds of acres of vacant land. This proposal has become a flashpoint for debates about the forms agriculture should take in inner cities. Proponents argue that urban farming must be scaled up to become a genuine engine of economic development. Detractors claim the Hantz project will result in large-scale industrial farming that exploits workers, sprays chemicals in neighborhoods, and does little for residents. They also point out that the Detroiters who have built a vibrant, indigenous urban-agriculture sector of gardens, farms, and associated community food projects have been taking care of vacant land in the absence of the sort of public support that Hantz is seeking for land assembly and preparation.

The city's planning staff has been grappling with the challenges posed by large-scale farming, vertical farming, and other proposals for untested forms of urban agriculture. They are also attuned to the broader potential of urban agriculture as an economic and community development, food-access, and vacant land–management tool. In August 2009, the city convened the Urban

Agriculture Workgroup, a stakeholder group charged with drafting a policy and zoning amendment to articulate the city's support for urban agriculture and to specify where and how food can be grown in Detroit.

In March 2010, the group reported back to the planning commission with a draft policy; meanwhile, commission staff is working on a zoning amendment that would define and permit a number of specific urban agriculture uses and activities. Despite Detroit's steady progress toward adopting an official policy and new regulations, however, the city faces a big obstacle from a surprising quarter: the state's Right to Farm Act.

As Kami Pothukuchi, associate professor of urban planning at Wayne State University, explains, Michigan law protects all commercial farming operations from nuisance claims as long as those farms comply with the state's Generally Accepted Agricultural Management Practices (GAAMPs). Although the act was clearly meant to protect rural farms from encroaching development, it does not define the term *commercial production* and contains no explicit exemptions for established urban areas. The danger for the City of Detroit is that as soon as it officially sanctions commercial farming through zoning, any associated development standards would automatically be preempted by state law. This could enable ventures like Hantz Farms to avoid local regulation. According to Rory Bolger, AICP, deputy director of the Detroit City Planning Commission, this threat of preemption has therefore stalled proposals to sell city-owned land to commercial agricultural operations.

The City has therefore requested that the state legislature amend the law to exclude established urban areas. Bolger remains optimistic about the prospects for action in the near future. "A policy for urban agriculture will emerge, and the framework has already been established," he says. "If the legislature addresses Right to Farm in early 2011, we could have a final policy and a zoning ordinance amendment by the end of the year."

Cleveland

While Detroit's attempt to turn large areas of vacant land over to productive agricultural use remains on hold, the city of Cleveland (pop. 444,313) is reimagining its built environment and embracing urban agriculture as an important and necessary part of vacant and abandoned property reuse. Problems of social inequity, chronic disease, obesity, and food deserts, along with an overabundance of abandoned property and vacant land, have created the perfect conditions for citywide urban-agriculture innovation and proliferation. With a common interest in and dedication to comprehensively transforming their community into "a cleaner, healthier, more beautiful and economically sound city," policy makers, local government agencies, nonprofit organizations, and the public are successfully collaborating to preserve and support opportunities for long-term community gardens and commercial urban agriculture throughout the city. (See City of Cleveland, City Planning Commission 2008.)

Within a five-year period, Cleveland has reformed its zoning code to establish an urban garden zoning district; passed an ordinance to permit the keeping of poultry, small farm animals, and bees throughout the city; implemented several training and grant programs to provide financial resources and education to beginning farmers; and initiated more than 100 new urban-agriculture projects. Cleveland's success in achieving these tasks is a result of multiple concurrent, dovetailing processes, including the formation of the Cleveland-Cuyahoga Food Policy Coalition and the development and implementation of a citywide sustainability plan, as well as strong agricultural roots, philanthropic foundation support, champions within local government, and the long-term commitment of several key organizations and individuals.

Urban and periurban agriculture in Cleveland dates back to the early 1900s. In 1907, the city established a program that set aside acres of public school land for horticulture education. While this program no longer exists, these spaces serve both as anchors for permanent urban-agriculture land use and reminders of agriculture's historical importance throughout the city. Many residents have strong agricultural roots, with parents or grandparents who grew up on rural farms and moved to the city in search of economic opportunities. "They brought with them not only an interest in agriculture, but agricultural knowledge," says Morgan Taggart, program specialist in agriculture and natural resources at the Ohio State University Extension Cuyahoga County (OSU Extension).

Since the 1970s, OSU Extension has promoted and supported community gardening and, more recently, commercial urban agriculture. In 2004, it collaborated with the Cleveland Department of Public Health (CDPH) to expand the development of community gardens throughout the city, particularly in underserved neighborhoods. With funding from the Centers for Disease Control and Prevention's Steps to a HealthierUS Program (now the Healthy Communities Program), OSU Extension and CDPH started more than 40 new gardens throughout the city and built new relationships with traditional and nontraditional partners, both individuals and organizations.

As a result of this work, in April 2007, OSU Extension and CDPH teamed up with Case Western Reserve University and the New Agrarian Center, a nonprofit organization that focuses on developing a sustainable local-food system in northeastern Ohio, to establish the Cleveland-Cuyahoga Food Policy Coalition. Coalition members represent more than 100 organizations, including city and county government, nonprofit and nongovernmental organizations, educational institutions, and private businesses, as well as farmers and producers. The coalition systematically addresses the region's food production, processing, distribution, access, health and nutrition, and disposal needs. Its five working groups—health and nutrition, community food assessment, land use and planning, local purchasing, and food waste recovery—provide the city and county governments with information and advice on the reform of existing policies and the creation of new policies and programs to support a health-promoting, sustainable, and community-based food system. The relationships formed within the Coalition provide a medium for the fertilization of new ideas related to urban agriculture.

Land-Use Regulatory Reform. In 2005, with funding from a USDA Community Food Projects Grant, OSU Extension developed a 12-week training program on small-business development for urban farmers. This program jump-started commercial agriculture in Cleveland by providing grants to assist farmers with start-up costs, but it also identified a significant barrier: regulations preventing the use of city-owned property for commercial agriculture, the sale of food grown on private or public property, and the keeping of chickens in the city. "The policies were a real barrier that prevented the increasing energy and interest in urban agriculture from moving forward," says Morgan Taggart. "The policies were antiquated."

The interest in commercial urban agriculture, combined with the need to productively reuse a growing number of vacant parcels, created the ideal conditions for policy reform. After conducting focus groups with urban farmers to better understand the issues and potential policy barriers, the coalition's Land Use Working Group (LUWG), consisting of several city and county planners, architects and urban designers, community development organizations, and a land trust, assisted city staff in a review of Cleveland's zoning code. "We picked out all the pieces that could be barriers, particularly the ordinances that could make it difficult to operate a community garden or an urban farm," says Taggart.

The Land Use Working Group worked closely with the Cleveland-Cuyahoga Food Policy Coalition and city planning staff to draft legislation for urban animal-keeping that addressed city council concerns about potential public and environmental health risks.

Growing Power

While the city's open-space recreation zoning district already allowed community garden use, it did not provide long-term protection and exclusive (i.e., more than interim) use for urban gardening. One-third of Cleveland's urban agriculture projects are located on land-bank lots, where they are considered an interim use that could be displaced by redevelopment at any time. To provide stronger protections for urban agriculture, the Cleveland City Planning Commission and the LUWG—with strong encouragement from Councilman Joe Cimperman and the leadership of City Planning Director Bob Brown—developed regulations for an urban-garden zoning district.

The new regulations create a zoning district solely for urban agriculture use. The district permits community gardens and market gardens (small commercial enterprises) and includes specific allowances for accessory structures and onsite sales, giving the city the ability to "reserve land for garden use through zoning." (See http://planning.city.cleveland.oh.us/zoning/pdf/AgricultureOpenSpace Summary.pdf.) As a result, land tenure has improved: to rezone sites now zoned as "urban gardens" requires public notice and public hearings.

"In the past, even on publicly owned properties, we could grant permission to garden or farm on a specific property, but then could turn around the next year and sell it for another use. Now in order to rezone property that is zoned for urban gardens for another land use, such as residential housing, you will need to go through a public process," explains Brown.

According to Taggart, political support and leadership by planning staff were key elements in the development of the new policy and its speedy adoption by the city council. Both Cimperman and Brown quickly embraced urban agriculture as an important emerging local land use and acknowledged that in some cases urban agriculture is the highest and best use of land. While the city continues to struggle with land-tenure issues for urban agriculture, this policy success created momentum for subsequent policy changes.

Shortly after the adoption of the urban-garden zoning district, the LUWG approached the planning commission with another policy issue: the keeping of poultry, small farm animals, and bees within the city. After cock-fighting problems in residential areas in 2005 and 2006, subsequent zoning-code amendments placed a six-chicken limit on residential parcels and required 100-foot setbacks for coops, effectively making it impossible to raise chickens or other small animals, such as rabbits, on small urban lots. The LUWG worked closely with the CDPH and city planning staff to draft legislation for urban animal-keeping that addressed city council concerns about potential public and environmental health risks, including concerns related to nuisances, injuries, illness, and vermin.

After a year-long revision and approval process, the ordinance now allows the keeping of poultry, livestock, and bees, subject to restrictions. Residents may keep up to six chickens, ducks, or rabbits and two beehives on a standard urban lot; coops and cages are restricted to rear yards with setbacks of at least five feet from side lot lines and 18 inches from rear lot lines. Larger lots and 100-foot setbacks are required for keeping roosters, turkeys, geese, goats, pigs, and sheep, and owners must license farm animals through the department of public health (City of Cleveland 2010).

The CDPH's involvement eased public and political concern about potential nuisance issues and contributed to quick approval and adoption of the ordinance by city council. Also instrumental was the political leadership of Councilman Cimperman. "He has supported the development of our innovative code to encourage urban agriculture uses and thinks about long-term land tenure for these uses," explains Taggart. Cimperman readily promotes the connections between the food system and quality of life, particularly health and nutrition: "We need to consider nutrition and health, the 20-year life expectancy difference between our white and black populations, when determining the highest and best use for land in our city."

Sustainability Planning. Between 1950 and 2009, Cleveland's population declined by 52.8 percent. As a result, Cleveland has approximately 3,300 acres of vacant land (of about 50,000 acres total) and 15,000 vacant buildings. While this is a formidable obstacle to economic and community development, local public and private sectors, nonprofit organizations, and residents have nevertheless embraced it as an opportunity to "re-imagine" their city. Vacant land is seen as a valuable resource that will enable Cleveland to "advance a larger, comprehensive sustainability strategy for the city, benefit low-income and underemployed residents, enhance the quality of neighborhood life, create prosperity in the city and help address climate change" (City of Cleveland City Planning Commission 2008).

With support and funding from the Surdna Foundation, the neighborhood development nonprofit Neighborhood Progress, Inc. (NPI; http://neighborhood progress.org) collaborated with the City of Cleveland and Kent State University's Cleveland Urban Design Collaborative (UDC) to initiate a sustainability planning process, beginning with a citywide study of potential innovative strategies for returning vacant land and buildings to productive use. NPI convened a

atives from the planning
nt, community develop-
various community and
d, Green City Blue Lake
The study mapped fea-
posed parks, greenways,
ead contamination, food
their locations to existing
rategies for the produc-
b loss and the resulting
artment or even the city
chtell, NPI's senior vice
strategies—both small
ated excitement among

de sustainability plan.
nable Cleveland (Figure
December 2008. The
goals, principles, and

The goals of Re-Imagining a More Sustainable Cleveland *were to find value in the city's growing vacant-property inventory and promote its strategic reuse, link the city's natural and built systems, and increase local food and energy self-reliance.*

Figure 4.1

strategies, including policy recommendations for returning vacant land and properties to productive use.

The plan's goals were to find value in the city's growing vacant-property inventory and promote its strategic reuse, link the city's natural and built systems, and increase local food and energy self-reliance. Strategies focused on neighborhood stabilization, green infrastructure development, and the integration of productive landscapes as an economic development strategy (CUDC 2008). The plan embraces urban agriculture, identifying community gardens, market gardens, and commercial farming operations as key productive landscape-reuse strategies, and outlines specific criteria for these uses, setting a goal of establishing a community garden within a quarter- to half-mile radius of every city resident (CUDC 2008, 32; see Figure 4.2).

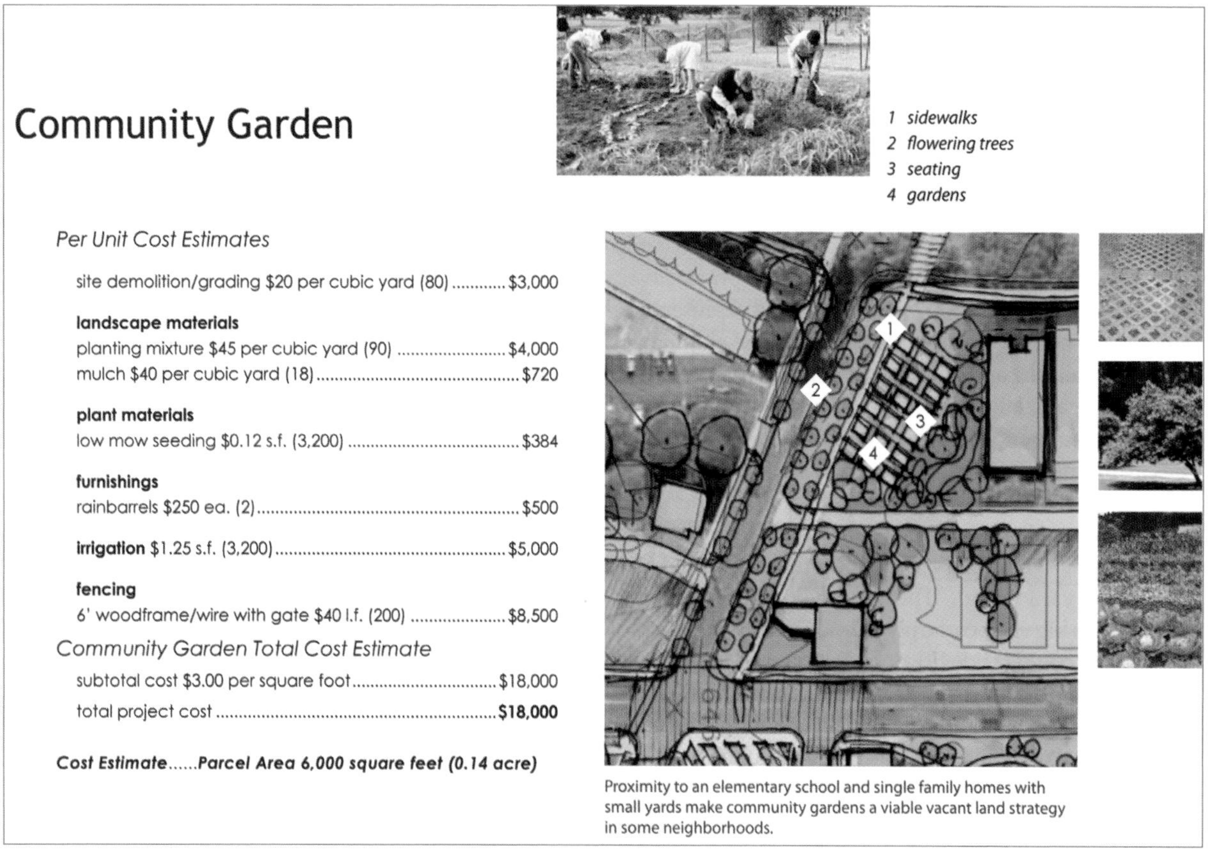

Community Garden

1 sidewalks
2 flowering trees
3 seating
4 gardens

Per Unit Cost Estimates

site demolition/grading $20 per cubic yard (80) $3,000

landscape materials
planting mixture $45 per cubic yard (90) $4,000
mulch $40 per cubic yard (18)..$720

plant materials
low mow seeding $0.12 s.f. (3,200) $384

furnishings
rainbarrels $250 ea. (2)..$500

irrigation $1.25 s.f. (3,200)..$5,000

fencing
6' woodframe/wire with gate $40 l.f. (200) $8,500

Community Garden Total Cost Estimate
subtotal cost $3.00 per square foot................................$18,000
total project cost **$18,000**

Cost Estimate......Parcel Area 6,000 square feet (0.14 acre)

Proximity to an elementary school and single family homes with small yards make community gardens a viable vacant land strategy in some neighborhoods.

Cleveland Urban Design Collaborative, Kent State University

Figure 4.2. One of Re-Imagining Cleveland's urban agriculture patterns

Additional policy recommendations promoting urban agriculture reuse of vacant properties include:

• Establish a task force to assess and address barriers to new vacant-land reutilization strategies, including zoning, building, and health codes, access to city land and water, and so on;

• Develop more detailed, parcel-based mapping of environmental contamination that distinguishes highly contaminated sites from less contaminated ones; include this information in the city's GIS parcel data;

• Develop parcel-level mapping of sites where children have tested positive for elevated blood-lead levels and factor this information into decision-making on building demolition in areas that have urban agriculture potential;

• Provide permanent support for local food production;

- Integrate permanent garden space in model block and neighborhood planning;

- Establish strategies for controlling use and new models for holding land (e.g., rezone to urban garden district, transfer ownership of land to community land trust, establish long-term land leasing with the ability to fence and secure);

- Develop policies and practices within the Cleveland Water Department that streamline farmers' and gardeners' access to water. Establish water rates that promote agricultural uses;

- Explore new ways of bringing water to sites, including maximizing the use of rainwater runoff from adjacent building roofs, leaving water lines to properties after demolition of buildings, and so on;

- Explore the potential for a municipal composting facility and community composting projects. (CUDC 2008, 31–32)

After the plan was adopted, NPI raised the financial resources to initiate a pilot grassroots reuse program in May 2009. The 56 projects that received funding (primarily HUD Neighborhood Stabilization Program funds and private foundation support) included 13 community gardens, 12 market gardens, three vineyards, and two orchards. "There were other proposals related to other reuse options, but a little more than half were related to food production," says Reichtell. These pilot projects will reuse 15 of the city's 3,300 acres of vacant property. Lessons learned will provide important information for future larger-scale projects.

"We are creating a movement. We are empowering people to look at the other vacant spaces in their neighborhood. This is not going to be solved with grant money. It's going to require people to take ownership and these pilot projects have made headway from an organizing perspective," explains Reichtell.

REUSING BROWNFIELDS FOR URBAN AGRICULTURE

As discussed in Chapter 2, brownfield sites present redevelopment opportunities and can be used for agriculture-related purposes. However, cities are wary of the potential health and environmental risks associated with on-site contamination.

With the exception of Kansas City, Missouri, none of the cities studied here has a full-fledged brownfields reutilization program, systematic testing for soil contamination, or clear standards for safe agricultural production on brownfield sites. Though some cities require raised-bed gardening on suspected brownfield sites, they do not specify particular techniques, materials, and procedures to minimize risk. In a few instances, local governments have expressed reluctance to even attempt such programs, due to food-safety and liability concerns. Risk-based assessment and redevelopment guidance from the U.S. EPA for urban agriculture on brownfield sites is needed to give state and local governments greater confidence in recommending remediation measures. Risk-based assessment and redevelopment standards tailor remediation strategies to the end use, such as residential, commercial, or passive recreation (Hollander et al. 2010).

Although guidance and standards are not yet developed, in early October 2010, U.S. EPA began to explore brownfields reuse for urban agriculture in a two-part webinar series. (See www.epa.gov/brownfields/urbanag/present.htm.) The webinars explored evidence-based questions such as:

Though some cities require raised-bed gardening on suspected brownfield sites, they do not specify particular techniques, materials, and procedures to minimize risk.

Nevin Cohen

- What are the common contaminants found in the air, water, and soil of an urban infill lot?

- How does this affect the decision process when developing an urban agriculture project?

- What kinds of historical land uses should trigger deeper investigation?

- Can amending urban soils with compost or other amendments make the soils "healthy"?

In addition, the webinars explored a range of policy questions surrounding brownfields and urban agriculture, including:

- How do existing state voluntary clean-up programs address urban agriculture issues?

- How does urban agriculture fit into existing programs for land-use determination?

- What are the property control and ownership considerations for urban agriculture projects?

- How should economic development drivers influence decisions for using urban agriculture as a revitalization strategy?

- How are innovative cities working within or changing their policy structures to make urban agriculture work for them?

The information in the webinars became the basis for more detailed conversations held at the Brownfields and Urban Agriculture Reuse Midwest Summit in Chicago in October 2010. Results of the summit will be made publicly available in 2011. The U.S. EPA website also includes several fact sheets on starting urban agriculture projects, as well as weblinks to success stories and other resources.

Despite the challenges outlined in the questions above, brownfields offer many opportunities for community transformation. This section provides an overview of how Milwaukee, Kansas City, Cleveland, and Chicago are developing and implementing programs and initiatives that encourage brownfield redevelopment for agriculture-related uses. It also explores some of the challenges faced by Philadelphia and Toronto in establishing such programs.

iStockphoto.com/BanksPhotos

Milwaukee

In Milwaukee (pop. 605,013), local government and the urban agriculture community are working with U.S. EPA Region 5 staff to reuse brownfield sites for urban agriculture. U.S. EPA has designated Milwaukee and nine other Environmental Justice Showcase Communities across the United States, committing $100,000 per city over a two-year period for demonstration projects that alleviate environmental and human health challenges. (See www.epa.gov/compliance/ ej/grants/ej-showcase.html.) Milwaukee's pilot project in the 30th Street Industrial Corridor focuses on redevelopment, green infrastructure mechanisms for stormwater management, and urban agriculture. The project builds on the EPA's significant brownfield program investments that have supported city and state commitments to address environmental and economic needs in this area. The project also includes actions to address public health issues and promote resilient neighborhoods and communities.

In Milwaukee, city officials have established an informal procedure for the reuse of brownfields as urban agriculture sites. The Redevelopment Authority of the City of Milwaukee (RACM) and the Department of City Development (DCD) work with local nonprofits and community organizations to first determine whether a community is particularly interested in a specific site for community gardening before assessing whether the property is a good fit for urban agriculture—meaning primarily that it is not a likely candidate for a property tax–generating use. In the 30th Street Corridor the emphasis has been on larger-scale sites, rather than small residential parcels, as candidates for remediation and urban agriculture. When a site is identified as having an interested community as well as low development potential, then RACM's brownfields planners undertake the brownfields assessment and remediation process.

Funds from EPA, the Wisconsin Department of Natural Resources (WIDNR), which regulates brownfields in the state, and some local grants are available for assessment. Traditional remediation for many sites is complete excavation of the contamination source, but this can be an expensive process; the city recently spent $100,000 to remediate a 0.7-acre parking lot across from an old paint factory to a clean-soil depth of 24 inches. Alternatively, a clay cap can be placed and clean material brought in to create a separation between contaminated and clean soil.

For urban agriculture on smaller sites or individual parcels, the city requires raised-bed construction and a clean-soil depth of at least 12 inches, a commonsense approach given its policy of neither conducting nor permitting soil testing on city-owned lots for liability reasons. While the city and WIDNR have no specific policies related to site remediation, RACM and DCD are beginning to explore the remediation of sites through added compost or other soil amendments. Remediation research has recently begun on two vacant, formerly residential parcels in the Lindsay Heights neighborhood through a partnership between Walnut Way Conservation Corp., a grassroots neighborhood organization focused on civic engagement, environmental stewardship, and economic enterprise, and the University of Wisconsin–Madison's College of Agriculture and Life Sciences. Walnut Way neighborhood residents, many of whom have a strong interest in urban agriculture, will be trained as "citizen scientists" to conduct research related to lead remediation, phytoremediation, and reuse of graywater for urban agriculture on the Lindsay Heights parcels.

Milwaukeeans have also begun to repurpose vacant industrial properties for aquaponics, an agriculture system that combines plant growing and fish production in a symbiotic relationship. Sweet Water Organics, a volunteer- and community-supported urban fish and vegetable farm (http://sweetwater-organic.com), has been in operation since 2008 in an old warehouse in the Bayview neighborhood. Inspired by Will Allen's Growing Power, Sweet Water's commercial-scale, sustainable aquaponics systems combine plant growing and fish production in a symbiotic relationship, producing a variety of vegetables, such as lettuce, tomatoes, basil, watercress, peppers, chard, and spinach, and also growing lake perch. The plants serve as a water filter in one tier of the recirculating systems, while perch grown in other tiers provide the waste that is a natural fertilizer for the plants. Outside the warehouse, greenhouses and compost piles have been built on the old parking lot. Sweet Water's for-profit enterprise partners with its educational foundation that teaches aquaponics, vermicomposting, and food production to children and adults.

Kansas City

Public officials in Kansas City, Missouri (pop. 447,306), actively link urban farming with the reutilization of brownfield sites. The city's brownfields initiative (KCBI) is housed within the City Planning and Development Department. Although the city has no policy requiring brownfields testing, KCBI provides such assistance upon request and works with neighborhood groups and other local organizations to develop strategies for crop production on brownfield sites. A partnership between KCBI and Kansas State University, supported in part by U.S. EPA grant funds, has provided training and technical assistance in brownfields redevelopment to Kansas City neighborhood organizations since 1998.

The U.S. EPA recently awarded the city and surrounding Jackson County a three-year, $1 million grant to conduct an assessment of hazardous substance and petroleum brownfields sites. The grant is managed by Kansas State University (which itself is conducting EPA-funded research on vegetable production on brownfields) in partnership with the Kansas City Center for Urban Agriculture, and includes funding for environmental assessments to identify possible sites for urban farms and community gardens. In 2010, KCBI submitted a second proposal to the EPA requesting funding for a smaller, 18-month project, a Sustainable Reuse Master Plan assessing the viability of urban agriculture for KCMO's former Missouri Correctional Institution. The site of the city's 327-acre Municipal Farm from the 1920s through the 1940s, the property is adjacent to a former tuberculosis sanitarium. Plans for the site include a one-acre community garden for 2011, with row cropping and ADA-accessible raised beds to be developed in future years.

Cleveland

Soil contamination is a significant problem in Cleveland. While all community garden and urban-agriculture pilot sites are required to be tested for lead and a few additional heavy-metal contaminants, the city does not have a method for systematically testing all vacant property for a comprehensive range of contaminants, including organic compounds. However, the U.S. EPA recently committed $100,000 to conduct a large-scale residential soil-contamination study across the city. Sites were prioritized for study based on land use, soil disturbance, and level of children's exposure to them, so school gardens—agricultural sites with the highest degrees of repeated soil disturbance and child involvement—were highest on the list. Sites were further prioritized by the expected contamination of adjacent sites.

Chicago

In Chicago (pop. 2,833,321), the city's long industrial past complicates its support for urban agriculture. In 1996, the City of Chicago, the Chicago Park District, and the Cook County Forest Preserve established the nonprofit land trust NeighborSpace to help community-based organizations protect community gardens and parks from development. Funded equally by local government and private sources, NeighborSpace acquires property from the city or other public or private owners and enters into long-term management agreements with community groups eager to garden on the land, ensuring public access and insulating users from potential liability. NeighborSpace currently owns 57 sites being used for community gardens and other urban agriculture projects. According to Kathy Dickhut, deputy commissioner of the Department of Zoning and Land Use Planning, the city wants to make sure that any public land that is being transferred to NeighborSpace for use as urban agriculture is tested for contamination and managed well to avoid health risks and minimize liability.

Representatives of the city's Department of Zoning and Land Use Planning and Department of the Environment say that the city wants to encourage urban agriculture in appropriate locations but also wants to be sure that food production is done safely and that all urban agriculture projects receiving assistance from the city support multiple goals. While city officials see urban agriculture as one piece of the local food access and security puzzle, they do not see a role for large-scale commercial farming operations in Chicago.

Chicago wants to encourage urban agriculture in appropriate locations but also wants to be sure that food production is done safely and that all urban agriculture projects receiving assistance from the city support multiple goals.

As Aaron Durnbaugh, deputy commissioner of the Department of Environment, explains, "If you're just trying to produce food, existing farmland and smarter farmland practices that keep soil on the farms are where to concentrate. I think urban agriculture needs to be tied into public health, educational opportunities, and recreational opportunities to be economically feasible, and not tied to the … vision of growing all of your food in the city."

Both Dickhut and Durnbaugh question the financial feasibility of using brownfields sites for interim-use urban agriculture. Site investigation and preparation are costly, and, as noted, there is currently no accepted standard for remediation when the end use is urban agriculture. To help remedy this, the city is developing protocols for growing food on land that the city owns or is transferring, selling, leasing, or designating for temporary uses. The draft recommendations require Phase I site assessments for agriculture projects on city-owned land, and both Phase I and Phase II assessments where the city is selling, transferring, or leasing land.[3] In both cases, sites would need barriers, such as existing concrete, a clay cap, or a rubber mat membrane, to isolate any growing medium from potential contaminants.

Philadelphia

Although Philadelphia has a long history of using greening and gardening as vacant-land management strategies, and though there are urban agriculture projects operating on brownfields sites in it, the city has not tried to integrate these efforts into a formal brownfields-redevelopment strategy, nor has it formulated an explicit policy on this practice.

Philadelphia's industrial history has left a legacy of contaminated soil, so the city encourages container farming in raised beds. According to Sarah Wu of the Mayor's Office of Sustainability, "The land is not a resource for soil but for open space." As Zoning Commission executive director Eva Gladstein explains, "You have to remediate to the highest level when you put food production on a brownfield." Because cleanup is cost prohibitive, the projects located on brownfields play a questionable role in any long-term land-recycling strategy.

The issue of regionalizing the food system is becoming more pressing as planners and policy makers confront rising energy prices and climate change.

Toronto

In Toronto, existing brownfields initiatives have yet to include urban agriculture as a use for contaminated sites. The city's Environmental Protection Office, housed within Toronto Public Health, is now developing a soil-contaminant testing protocol to assess potential risks for different land uses. The protocol will be used to gauge the suitability of a parcel for urban agriculture.

ECONOMIC DEVELOPMENT

Although many older industrial cities, as well as those with dense, built-out areas, remain reluctant to tackle brownfields remediation and reuse for urban agriculture, it behooves cities to make careful assessments of the development potential of these sites. In order to revitalize communities, it is becoming increasingly clear that "highest and best" economic uses will not be found for every vacant or underused property. As interest in urban agriculture continues to grow, planners will need to understand more fully its potential as an economic sector. At present, this is the least-documented aspect of urban agriculture.

According to the U.S. Department of Agriculture's Economic Research Service, in 2009 U.S. residents spent more than $600 billion on food prepared at home and more than $526 billion on food purchased outside the home. This presents a significant economic opportunity for regional food systems that wish to tap into this revenue stream. The issue of regionalizing the food system is becoming more pressing as planners and policy makers confront

Nevin Cohen

rising energy prices and climate change. Reliance on foods from Europe, Asia, and South America—or even distant regions in the United States—is cost-effective for households as long as energy prices remain relatively cheap and water is readily available. Many urban gardeners and farmers have recognized these trends and argue that the survival of our cities depends on bringing food production closer to home. "When you look at the food needs of communities," says Harry Rhodes of Growing Home in Chicago, "the natural answer is to be growing food in cities."

Indeed, many urbanites are gravitating toward this answer. The American Community Gardening Association estimates that there are more than 18,000 community gardens in the United States and Canada. In the last three years,

Detroit has seen a tremendous growth in the number of family, school, and community gardens, from 41 family gardens to 220, and 39 community and school gardens to 134 (Whitesall 2007). Milwaukee's Growing Power trains 3,000 aspiring urban farmers each year—and is planning to build a five-story vertical agriculture building for year-round vegetable production.

Many economic development planners assume these activities are too small to matter. However, a recent study by Michigan State University estimates that by repurposing Detroit's vacant land into urban farms, community gardens, storage facilities, and hoop houses could supply Detroit residents with more than 75 percent of their vegetables and more than 40 percent of their fruits (Michigan State University 2010). The small size of urban farms and gardens allow them to start up (or wind down) production relatively rapidly in response to changing market conditions.

Moreover, planners and economic development officials find that traditional development tools have been seriously diminished by the faltering economy—few housing developments or factories are now being built. In light of this, some cities are discovering that cultivating clusters of local food businesses can be a potent way to create stronger local economies. Such work can penetrate to populations that have been overlooked by traditional development strategies.

In Flint, Michigan, where the erosion of the auto industry has vacated thousands of acres of factories and obliterated 16,000 jobs, the Genesee County Land Bank took over abandoned homes, clearing the land when needed and restoring homes when practical. This land trust now owns one-fifth of the city's land base, approximately 5,600 properties in the Flint area. For a reasonable rate they sell the land to people wishing to grow food. Two karate instructors, Jacky and Dora King, purchased over 20 acres of this vacant land to create Harvesting Earth Educational Farm, a nonprofit farm that teaches youth about farming and karate. Jacky built a hoop house (a simple metal-framed greenhouse) and invited nearby youth to help him farm, promising them both job skills and solid experience to show to employers at the end of their tenures. Youths have pitched in eagerly, selling produce to earn part of their incomes (Michigan Municipal League 2010). A sister farm, Flint River Farm, hopes to turn a polluted site where the city has dumped leaves for more than a decade into a thriving compost business and working farm. With support from community partners such as the Ruth Mott Foundation, these initiatives have helped persuade the city to consider food production a civic goal.

Flint's urban farmers maintain that the 12,000 vacant lots—with existing water mains—should be considered arable farmland. Community leaders now point to the possibility that rust belt cities such as Cleveland, Detroit, and Toledo have so many acres available they could support not only food production but also manufacturing firms creating the inputs or tools urban farmers need, as well as food-distribution businesses linking urban farms to nearby consumers.

Already, The Greening of Detroit (www.greeningofdetroit.com), a nonprofit organization dedicated to creating a greener Detroit through planting and educational programs, environmental leadership, advocacy, and building community capacity, has formed a cooperative distribution service that conveys $480,000 of foods annually raised on community gardens to urban food buyers. Pollo de Campo, a Latino/Anglo poultry co-op in periurban Minneapolis, raises free-range chickens on aggregated quarter-acre sites, using a model that could easily be adapted to urban lots (Haslett-Marroquin 2009). Janus Gardens in Portland, Oregon, has trained inner-city housing-project residents to start their own farms. Flats Mentor Farms, a 70-acre farm in Lancaster, Massachusetts, trains experienced Hmong, Kenyan, and

Liberian farmers to do business in their new country. Greensgrow Farms took over a one-acre brownfield in Philadelphia and now farms it organically, selling greens and vegetables to local restaurants and "city-supported agriculture" shares to the community. In 2009, their annual gross nursery and farm sales totaled $1 million (Breaking Through Concrete 2010b). A comprehensive survey of community and squatter gardens in Philadelphia estimated that in 2008 gardeners produced food with a retail value of $4.9 million (Vitiello and Nairn 2009).

Chicago's Growing Home urban farm finds that farming instills entrepreneurial skills in inner-city residents. About 65 percent of program participants find subsequent employment or educational training, and 90 percent find stable housing after completing Growing Home's jobs training program (Breaking Through Concrete 2010a). In 2008, Minneapolis's Institute for Agriculture and Trade Policy launched its Mini Farmers Market project to increase access to fresh foods in local neighborhoods. Located at community centers, at churches, and on busy street corners, these producer-only markets of five or fewer vendors provide new opportunities for small farmers to increase their product sales. Minimarkets are eligible for relaxed permitting processes as well as reduced licensing fees. In the first year of the program, six new minimarkets were added; by 2010, the number had increased to 21. In at least one instance, the minimarket project has offered youth from community-development and job-training programs the opportunity to match their skills in food production with entrepreneurial skills.

While each of these programs is relatively small-scale, economic multiplier effects such as sales, earnings, tax revenue, and jobs become evident as producers become networked together. Planners should work to foster investment in the physical and knowledge infrastructure that will nurture urban farms and help them thrive.

Planners have a significant role to play in making sure that urban agriculture is part of planning for healthy neighborhoods.

COMMUNITY HEALTH AND WELLNESS

A diet rich in fruits and vegetables is associated with positive growth and development, improved weight management, and decreased risk for chronic disease (Gustafson, Cavallo, et al. 2007; Rolls, Ello-Martin, et al. 2004; U.S. HHS and USDA 2005). However, most Americans do not meet federal health guidelines for fruit and vegetable consumption. Four of the six leading causes of death in the United States—heart disease, stroke, diabetes, and certain cancers—are diet-related chronic diseases, and rates of

Kimberley Hodgson

overweight and obesity, which are risk factors for these diseases, continue to increase—particularly in minority and low-income populations (Levi, Vintner, et al. 2010).[4] Obesity and related chronic diseases pose a serious burden on the physical and financial health of individuals, businesses, and communities in the United States, costing more than $117 billion annually in forgone wages and costs of treatment (U.S. HHS–ASPE 2003). While access to safe and nutritious food is considered a basic right by the World Health Organization and the United Nations (World Food Summit 1996), the most affordable and accessible foods for many children and adults, particularly in low-income households in both rural and urban areas, are calorie-dense but nutrient-poor. This is in part due to easier access to fast-food restaurants and convenience stores than to healthful food sources (also referred to as "food swamps"; Rose et al. 2009) coupled with a lack of supermarkets or other sources of fruits and vegetables such as farmers markets (Sallis and Glanz 2006; Powell, Slater, et al. 2007a, 2007b; Babey et al. 2008; Sallis et al. 1986; Cheadle et al. 1993; Horowitz et al. 2004).

Planners and public health practitioners increasingly recognize that the quality of the built environment plays a determinative role in the health of individuals and neighborhoods (Northridge, Sclar, and Biswas 2003). Planners have a significant role to play in making sure that urban agriculture is part of planning for healthy neighborhoods. In two of the cities studied, Toronto and Minneapolis, city departments of health initiated local food-systems and urban-agriculture work and fostered coordination and cooperation among other city departments, including planning. A third city and region, Seattle/King County, is building upon its long tradition of community gardening to more broadly embrace urban agriculture and its public health benefits.

Toronto Public Health has now placed urban agriculture within a citywide Food Strategy project.

Toronto

Toronto Public Health, the agency that created the Toronto Food Policy Council in 1991, has now placed urban agriculture within a citywide Food Strategy project, a set of action steps to create a "health-focused food system" (City of Toronto, Medical Officer of Health 2010). A 2008 scan of city agencies revealed a range of food-related objectives and asked how these various food initiatives, including urban agriculture, could be effectively united as a focused health strategy, given rising health care costs. The resulting May 2010 report, "Cultivating Food Connections," notes the work of the Toronto Environment Office (above) as an example of positive government action to create "food friendly neighborhoods" built on city-community

Cultivating Food Connections:

Toward a Healthy and Sustainable Food System for Toronto

May 2010

TORONTO PublicHealth

linkages. Toronto Public Health staff are responsible for coordinating and implementing the food strategy recommendations, which include expanding the city's financial support of urban agriculture projects ($800,000 in 2008 and 2009), as well as identifying neighborhoods in need of better access to food and targeting them for additional community gardens and food production spaces.

Seattle/King County

The City of Seattle and King County have long been hotbeds of progressive action, and they are likewise on the cutting edge of the urban agriculture movement. Seattle is widely recognized as a national leader in community gardening, but a recent groundswell of interest in commercial farming and entrepreneurial agriculture goes beyond support for community gardens. This is attributable in part to a popular awareness of food security and public health aspects of local food systems.

National interest in Seattle's urban agriculture leadership has focused on the city-run P-Patch community gardening program (www.seattle.gov/ neighborhoods/ppatch). Housed in Seattle's Department of Neighborhoods, P-Patch has been a model for cities across the country for almost 40 years. Currently, the program includes 73 gardens serving more than 2,000 households on 23 acres of city land. According to Erin MacDougall, Healthy Eating and Active Living program manager for Seattle/King County Public Health and a member of the volunteer board guiding the nonprofit P-Patch Trust (www.ppatchtrust.org), Seattle's community gardening program is so popular that more than 1,200 people are on the wait list for new plots. "People are spending anywhere from a year and half to three years on the waiting list," says MacDougall. "There's a very high demand in the city for a very small 10-by-10 P-Patch plot."

As University of Washington planning professor Branden Born sees it, the logjam at P-Patch has inspired the city to reorient its thinking about urban agriculture and recognize its diversity of expression in the area. Beyond P-Patch, there are numerous groups in Seattle and King County pursuing or encouraging a variety of urban agriculture models. These groups include nonprofit, cooperative, and commercial growers as well as government agencies, public universities, and other organizations offering land, education, or technical assistance.

For example, the nonprofit Alley Cat Acres Urban Farm Collective (www .alleycatacres.com) is building a network of small farms on vacant or underutilized land to provide affordable fresh food to underserved families, while City Fruit (http://cityfruit.org) works with neighborhood residents to grow healthy fruit, harvest and use or put up what they can, and share the surplus with others. On the public side, the University of Washington operates a farm as an educational tool to keep the university community in touch with how food is grown, and King County hosts three gardens in its park system. Additionally, there are small businesses such as the Seattle Urban Farm Company that provide goods, technical assistance, or training to help residents grow food on private land.

Overall, the urban agriculture community in Seattle and King County is more cohesive than in many other leading places for urban food production. Practitioners and advocates have used both formal mechanisms such as the Seattle Good Food Network and less formal mechanisms such as the Urban Farm Hub blog to communicate and collaborate. (See www .urbanfarmhub.org.)

The Seattle Good Food Network originated in a cooperative effort among King County, Washington State University–King County Extension, and the Washington State Department of Agriculture to convene a diverse group of stakeholders in order to take a comprehensive look at food-system issues. This group met regularly as the Acting Food Policy Council from 2006 to 2009 but was never officially sanctioned by any unit of local government. Upon the Puget Sound Regional Council's formal approval of a regional food policy council in 2009, much of the acting council's energy was transferred to the Seattle Good Food Network, coordinated by Born.

From a planning and policy perspective, Seattle's 2007 comprehensive plan, which is focused on sustainability, makes a number of references to community gardening. For example, the urban village element appendix states a goal of providing at least one community garden for every 2,500 households in the city's designated urban villages (City of Seattle 2005/ 2007). Although King County's most recent comprehensive plan does not specifically mention urban agriculture, it does include a number of policies supporting rural farmland, soil health, and aquatic food resources. According to MacDougall, much of the interest in urban agriculture in the county comes out of a friendly competition with Seattle and is generally not attributable to resident advocacy.

Much of the city's recent policy work related to urban agriculture can be traced to City Council president Richard Conlin and his Local Food Action Initiative (LFAI).[5] LFAI tasked various departments with specific actions related to food-systems policy and planning. For example, it asked the Department of Neighborhoods to draft a food-policy action plan and to identify infrastructure for urban agriculture. The resolution charged the Department of Planning and Development (DPD) with assessing regulatory barriers to urban agriculture, and it tasked the Office of Economic Development with assessing city policies that affect farmers markets and market

In Seattle/King County, internal policy changes are breaking down some of the barriers to expanding urban agriculture.

Kimberley Hodgson

gardens. Subsequently, internal policy changes are breaking down some of the barriers to expanding urban agriculture. For example, in 2009 the Department of Transportation eliminated the permitting fees for residents who want to grow food in right-of-way planting strips between the sidewalk and the roadway.

A month after Mike McGinn took office as mayor in January 2010, he and Conlin announced the 2010: Year of Urban Agriculture campaign to promote urban agriculture and access to local food (www.seattle.gov/urbanagriculture). On the policy front, the most promising development is a zoning code revision that clarifies and broadens the city's support for food production and related activities. Andrea Petzel, a land-use planner with DPD who took the lead on the code reform project, says, "We didn't want urban agriculture to be relegated to the periurban areas or industrial lands or be something that was exclusively within the control of the city."

In August 2010, the city approved the zoning code revision, adding new definitions for urban farm and community garden, expanding allowances for farm animals, and including broad permissions for urban agriculture–related activities with limited permitting processes. (See City of Seattle 2010b.) For example, community gardens are now allowed by right in all zones, except for heavy industrial land, where they are allowed only on rooftops and the sides of buildings. The revision also permits urban farms, which allow the growing and selling of food on the same lot, in all zones, including residential ones.

Seattle's biggest barrier to expanding urban agriculture is land cost. Because the city has little vacant land and relatively few brownfields compared to places such as Philadelphia, Cleveland, or Detroit, there has been little discussion locally about the role urban agriculture might play in recycling abandoned or contaminated properties.

However, Born points out that places like Seattle have schools, utility corridors, and other pieces of land that could be co-used for urban agriculture. "There is real potential if we start thinking differently about urban agriculture in the city," he says. As Petzel sees it, "so much of the interest in urban agriculture is driven by community residents and organizations, and we have to make some fundamental changes to our city policies—be it the zoning changes or opening up municipal land for people to actually be able to grow and sell food. The challenge is coordinating the nuances as a city and being really crystal clear with the public."

Minneapolis

Minneapolis (pop. 386,691) is part of the seven-county Minneapolis–St. Paul metropolitan area renowned for its strong tradition of regional planning through the Metropolitan Planning Council, established by the Minnesota legislature in 1967. Although comprehensive planning is not mandated statewide, the Metropolitan Land Planning Act requires every local government in the region to develop a comprehensive plan. When the City of Minneapolis updated its comprehensive plan in 2009, it introduced urban agriculture into four chapters: open space and parks, the environment, public services and facilities, and urban design.

In keeping with its strong comprehensive-planning tradition, Minneapolis is also actively engaged in planning for sustainability. In 2003, the city council adopted a resolution initiating the Minneapolis Sustainability Program, and in 2005, the comprehensive plan was amended to include key sustainability indicators and mandate their use across all 18 city departments. These indicators were revised through a public process in 2009 to include local foods, waste reduction, and recycling.

With respect to urban agriculture, the Minneapolis–St. Paul metro area has a decades-long tradition of home gardening, master gardening (through the University of Minnesota Extension), and farmers markets. While the state contains multigenerational farming communities typical of the rural Upper Midwest, periurban Minneapolis–St. Paul is farmed by growers who rent land and produce food to sell at the city's farmers markets and upscale local restaurants. Many of these periurban farmers are Hmong immigrants, who brought a subsistence-farming culture with them to the United States.

Local food production is an important part of Twin Cities culture. Minneapolis is home to the largest concentration of natural-foods cooperatives in the country—and these co-ops have been at the core of the metro area's sustainable agriculture movement for the past 30 years. Over the past decade, a sizable community-gardening movement has also developed, with more than 200 community gardens in the metro area. In 2004, the McKnight Foundation funded a study that documented the challenges faced by local community gardens. The resulting *Twin Cities Community Garden Sustainabil-*

ity Plan (2005) called for a community garden association that could advocate for gardens and help community gardeners network, organize, and work collectively. Housed in its start-up phase at the nonprofit Green Institute, Gardening Matters is the independent community-garden organization now dedicated to that purpose.

In 2007, the Institute for Agriculture and Trade Policy (IATP) began the Minneapolis Mini Farmers Market Project described above. (See map, Figure 4.3.) The nonprofit IATP serves as the umbrella organization through which minimarkets can participate in the Farmers' Market Nutrition Program; it undertook a role that could be played by planners, working with the City of Minneapolis to simplify the permitting and licensing process for these markets. Correlated results include an increase in farmers markets around the city, as well as a 20 percent increase in vegetable consumption among

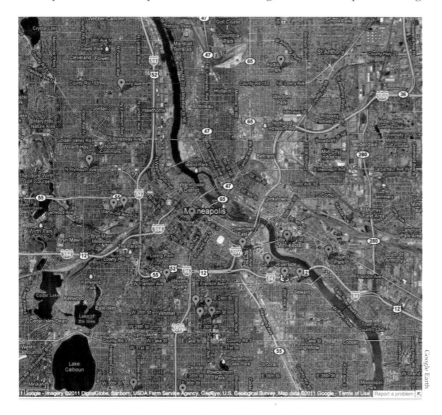

Figure 4.3. The Minneapolis Mini Farmers Markets, 2010

patrons at several of the minimarkets (IATP n.d.). This project, and other activities revolving around comprehensive planning, sustainability planning, and community gardening, set the stage for the large, multistakeholder planning process known as Homegrown Minneapolis (HM), which is creating the future of urban agriculture in that city.

Homegrown Minneapolis Phase One: The Process. Championed by Mayor R. T. Rybak, phase one of the Homegrown Minneapolis (HM) Initiative began in December 2008. HM was "built on the idea that a strong local food system can positively impact the health, food security, economy and environment of our city and the surrounding region" and that the city could "play an important role in this process by supporting residents' efforts to grow, sell, distribute, and consume more fresh, sustainably produced and locally grown foods" (MDHFS 2009, 1). The mayor designated the city's Department of Health and Family Support as the primary agency to provide staff support and coordination as part of a five-year federal grant targeted toward preventing obesity through increased consumption of healthy local foods. Two city council members have also been deeply involved in HM.

Between January and April 2009, more than 100 stakeholders representing the city, schools, parks, local businesses, neighborhood organizations, nonprofits, residents, and other organizations met regularly to discuss the strategic planning and collaboration needed to bring the idea underlying HM to life. According to Karin Berkholtz, city planning manager, who was heavily involved in phase one, the Minneapolis planning environment is extremely participatory: "The public's expectation is for a high degree of community engagement and their appetite for it is huge." The process was also intended to catalyze collaborative food-system activity, which at the outset of the process was not well organized.

HM contains many elements of a community food assessment. It identifies strengths and gaps in the local food system; a particular strength is the city's strong base of small-scale production and distribution of locally grown foods. The gaps are familiar across cities: inequitable access to healthy foods, the lack of small- and mid-sized infrastructure to support local food production and distribution, soil contamination and remediation issues, lack of communication and coordination among farmers markets, and a lack of connection between rural and periurban producers and urban consumers (MDHFS 2009).

At the outset, HM focused on four key areas: farmers markets; community, school, and home gardens; small-enterprise urban agriculture; and commercial use of local foods. The process resulted in 72 recommendations and 146 detailed action steps, including designation of the parties responsible for implementing them. Six key recommendations ranged from passing a city council resolution that would put support of healthy local food on record and create a work group to oversee HM's implementation to creating city policies and developing systems, tools, and a public education and communications campaign to support the local food system. In addition, local-foods jobs and small-enterprise urban agriculture will be included in the city's Green Jobs Initiative.

Homegrown Minneapolis Phase Two: Implementation. Since June 2009, when the Minneapolis City Council received the final HM report, seven work groups have been created to focus on implementation efforts; each work group has at least one city staff person assigned as a convenor, and city departments have been tasked with moving recommendations forward. (See www.ci.minneapolis.mn.us/dhfs/hgimpefforts.asp.) The council quickly took HM's first recommendation to heart and passed resolution 2009R-283, "Recognizing the Importance of Healthy, Sustainably Produced and Locally Grown Foods and Creating the Homegrown Minneapolis Implementation Task Force." (See Figure 4.4.)

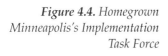

Figure 4.4. Homegrown Minneapolis's Implementation Task Force

Perhaps the strongest result of HM has been the unfolding implementation of the report's Recommendation 5: "Prioritize local food production and distribution when determining the highest and best use of City-owned and private land and when planning new development or re-development projects that could potentially affect existing local food resources" (MDHFS 2009, 12). This spurred the Department of Community and Economic Development to begin work on an urban agriculture policy plan for the city. According to Amanda Arnold, principal city planner, the policy plan will focus on eliminating zoning and land-use barriers to urban agriculture. (The current zoning ordinance permits community gardening in all but three zoning districts, but it does not address commercial growing.) It will identify the appropriate locations and needs for different types of urban agriculture activity across the city and provide supportive regulations and guidance.

Arnold notes that the urban agriculture policy plan process will be somewhat unusual in that it is an outgrowth of a preexisting, very large community engagement process. Urban agriculture stakeholders have already played significant roles in articulating the need for the plan during the first phase of HM, and Arnold expects that their participation will continue at similar levels. As of early December 2010, the policy plan is open for public review and comment until the end of January 2011. The urban agriculture policy plan is scheduled for consideration by the City Planning Commision at the end of February 2011, after which the comprehensive plan will be amended to include it (City of Minneapolis 2010b).

Urban agriculture stakeholders, described by one stakeholder as "a growing choir of voices," are playing a much greater role in Minneapolis city planning and policy making than they did before. Arnold says that phase one of HM helped develop a relationship between the urban agriculture community and local government that is still evolving, as are stakeholders' understandings of the issues and of one another. Nonprofit organizations look forward to meaningful collaboration as phase-one recommendations are implemented. For its part, the city expects that phase two of HM—and in particular the development of the urban agriculture policy plan—will help identify the "next round of champions" for urban agriculture in the community. The plan's adoption will by no means be the end of the process. Interest in an advisory Food Policy Council is strong, and advocates hope to have one permanently in place by the time phase two ends in the summer of 2011. In addition, the city is working to develop measurable local-food sustainability indicators to accurately track how much food is produced and consumed in Minneapolis.

Homegrown Minneapolis illustrates how, given a jump start from an urban agriculture champion (in this case Mayor Rybak), cities can develop consensus about urban agriculture and food system goals. The considerable political and public will mobilized through this participatory process has resulted in rapid progress toward further urban agriculture planning and implementation. Planners played a key role in the boundary spanning and bridge building that was fundamental to the process.

COMMUNITY CAPACITY BUILDING AND EMPOWERMENT

One of the defining characteristics of urban agriculture in North America—from the days of wartime victory gardens, through the community gardening movement of the 1970s, to today's urban agriculture coalitions—is the ability of urban gardeners and farmers to organize themselves to spur community empowerment and self-determination, notably in challenged urban communities. Urban agriculture, especially community gardening, has been shown to build community capital (human, social, political, and economic) such that, at its most effective, urban agriculture is as valuable for the organization and process by which food is produced as it is for the food itself (Smit and Bailkey 2006). Greening projects, especially those that revolve around community gardening and other forms of urban agriculture, can help communities organize and build their capacities and abilities to rebound from stressors and adapt to change—the hallmarks of resilience (Tidball and Krasny 2007; Dubbeling et al. 2009). These opportunities for collective organization can become especially valuable in socially marginalized communities and among new immigrant groups.

Greening projects can help communities organize and build their capacities and abilities to rebound from stressors and adapt to change.

Margrethe Horlyck-Romanovsky

Planners and other local government staff can play important roles in supporting grassroots groups and efforts involving urban agriculture. Community-based groups often have intimate understandings of neighborhood and individual-level issues, and thus can be resources for informal communication and learning between local governments and urban agriculture practitioners. In some cases, however, local government can be indifferent or even pose obstacles to the community capacity-building activities of grassroots urban-agriculture groups. The following section explores the role of such groups in New York City, Vancouver, Los Angeles, Kansas City, and New Orleans in building community capital, influencing planning strategies to integrate urban agriculture into the built environment, and contributing to broader goals of social sustainability.

New York City: South Bronx Casitas

The highly cosmopolitan atmosphere of New York City (pop. 8,391,881) offers numerous examples of how urban food production holds importance across different cultures, as many new Americans seek to maintain their agricultural skills in their new country. For example, scattered throughout the South Bronx are individual *casitas* ("little houses") built by Puerto Ricans that serve as both social centers and representations of the country villages left behind. Typically, each *casita* occupies one or more vacant parcels, contains a variety of individual and shared food-growing spaces, and is centered around a small, quasi-residential structure of traditional design and decor that serves as a clubhouse of sorts and hosts social activities. Each *casita* is a stable, collectively managed connection to Puerto Rican roots amid the demands of a new culture. One of the oldest and best known, Rincon Criollo (loosely translated as "down-home corner"), now occupies 3,000 square feet on two corner lots at East 157th Street and Brook Avenue, after a 2007 move from its original, city-owned site one block away. As part of the GreenThumb garden program of the city's parks and recreation department, which is intended to foster civic participation, spur neighborhood revitalization, and preserve open space. Rincon Criollo must be open to the public a minimum of 10 hours per week and conduct a variety of cultural and educational events. (See www.greenthumbnyc.org.) These requirements are meant to anchor it, as a community garden, to its immediate community. But it is equally important to see the *casitas* of the South Bronx as cultural anchors that are not only examples of self-organization around urban agriculture but models for the maintenance of cultural traditions around the growing and consumption of food.

Vancouver: Neighborhood Food Networks

Since 2008, Vancouver has experienced the emergence of the neighborhood food networks (NFN), partnerships among community groups, local health and housing office staff, and residents. An outgrowth of the Village Vancouver Transition Town Initiative, which seeks to create more resilient, complete, and low-carbon communities, the NFNs focus on the neighborhood food system, particularly the number and location of community gardens, small-scale food processors, and composting facilities.[6] Vancouver currently has seven NFNs, with several more in development. The NFN provides a medium for networking and collaborating; growing, sharing, and celebrating; and enabling people to come together in a neighborhood setting. Where the Vancouver Food Policy Council provides a formal relationship between city government and community groups, NFNs allow for less formal communication and learning. An advantage of the NFN model is that it promotes direct connections among residents and an intimate understanding of local issues.

Los Angeles: A Tale of Two Urban Farms

As in New York City, urban agriculture practice in Los Angeles (pop. 3,833,995) reflects immigrants' view that agriculture is a vital cultural practice. But two very different stories show local government's ambivalence toward this use.

Not far from the glamour of Hollywood Boulevard, the Wattles Farm and Neighborhood Garden occupies over four acres of highly valued real estate at the foot of the Hollywood Hills. Begun in 1972 and incorporated as a nonprofit organization six years later, the Wattles Garden accommodates more than 160 gardeners, many of them Russian immigrants, on land leased from the city's parks and recreation department (Lawson 2005). Over four decades, the gardeners have developed a form of self-management that calls for each of them to be responsible to the others, in terms of governance and the shared maintenance of the site. The gardeners must also meet the expectations stated in their cooperative lease agreement with the city, thus helping to ensure their long-term tenure on the land. Here, a mutually beneficial, long-term relationship exists between urban farmers and city government.

In another Los Angeles neighborhood, however, the South Central Farm (see Chapter 2) offers an example of less successful community organizing and less productive political relationships around urban agriculture. In 1980, the City of Los Angeles acquired by eminent domain 14 acres of industrial property south of downtown from nine private landowners, to use for a trash incinerator. Citizen protest eventually led the city to abandon its plans and set aside the land for a community garden as a positive response to the riots that followed the 1992 Rodney King court decision acquitting four Los Angeles Police Department officers of brutality. In 1994, title to the property was transferred to the Harbor Department, which contracted with the Los Angeles Regional Food Bank to continue operating the site as a community garden.

South Central Farm, called the largest urban farm in the United States with more than 350 low-income Latino immigrant families producing fruits and vegetables, operated in this fashion until August 2003, when the City Council in closed session approved the sale of the property to a local developer. (See South Central Farm 2005.) In an attempt to retain the use of the property, the Latino farmers immediately organized as South Central Farmers Feeding Families, but within months the developer issued a termination notice for the garden. The farmers sought and received a temporary injunction in Los Angeles Superior Court halting the process but were ultimately evicted in 2006 and dispersed to other sites. (See Figure 4.5.)

Farming the industrial site was inconsistent with the city's plans for the larger area; thus, despite the healthy food access it created for a large number of people, the farm was always a temporary land use, and its location adjacent to the African-American community of South Central Los Angeles provoked racial tensions among the farmers, many of whom lived outside of the city, and political and community leaders. The developer's decision to evict the farmers from the site mobilized them into political action, but the process exposed significant tensions within their organization over the generation of individual revenue, political engagement, and representation of interests.

Figure 4.5. South Central Farm farmers in Los Angeles stage a streetside protest following their eviction.

Martin Bailkey

Kansas City, Missouri: Amending Municipal Zoning Language

The Kansas City metropolitan area straddles the Kansas–Missouri border and sits at the eastern edge of the vast region that since the 1840s has been transformed from tallgrass prairie to agricultural land. The 1923 zoning ordinance in Kansas City, Missouri (KCMO) contained language permitting farming, greenhouses, nurseries, and truck gardening within single-family residential districts. In the 1960s, extensive land annexation by KCMO brought thousands of acres of working farmland north of the central city within municipal boundaries and placed them under agricultural zoning. The many municipal jurisdictions in the area—including KCMO, Kansas City, Kansas (KCK), and many suburbs in both states, as well as county governments—has resulted in a patchwork of laws and regulations affecting urban farming. But a zoning change approved by the KCMO City Council in June 2010 to permit market farming in residential districts is being watched closely—perhaps to be duplicated—by other governments in the region.

Kansas City's urban agriculture community mobilized into advocacy in late summer 2009, when KCMO code enforcement officials visited Badseed Farm, a successful three-acre CSA operating in a relatively affluent neighborhood on the city's south side. Prior to this, the expansion of city farms and CSAs in the region had elevated urban farming out of relative obscurity and into public view, thus exposing it to public opposition and the employment of various policy barriers. Revenue-generating CSA farms were noted as operating on the regulatory margins in KCMO, and governance problems arose because of a lack of definitions for urban agriculture types, such as community gardens and CSAs, in the city's zoning ordinance.

Badseed Farm (Figure 4.6; www.badseedfarm.com) did not appear to be a problem for the majority of its neighbors. However, one neighbor, the owner of a vacant residential property next door, felt that an adjacent working farm with chickens and goats threatened to lower the house's value. She called city inspectors, but they did not respond quickly. The issue received local media attention, the city council became involved, and soon the Badseed farmers were cited for violating restrictions on accessory use of a residential site and for employing paid interns as farm assistants, which was also restricted by residential zoning.

The KCMO zoning and development code had just been revamped earlier that year, but it did not fully address the various regulatory issues surrounding urban farms. This gave city farming advocates a clear focus for conversations with city officials that fall. A meeting that the advocates had with the city's planning director led to the formation of exploratory committees focusing on livestock keeping and appropriate code language for urban farming. In the meantime, the city council member whose district contained Badseed Farm became a champion of urban agriculture within KCMO government. He requested that the City Planning and Development Department shepherd the process, and he recruited cosponsors within the city council to begin working on an amendment to the zoning code. During the winter of 2009–2010, supporters developed an extensive advocacy network through e-mail lists and social networking sites, while the code language committee talked with numerous neighborhood organizations. Badseed Farm continues to operate and plans to expand to include an onsite farm market in 2011.

On June 10, 2010, the city council approved amendments to Chapter 88-312 (Agriculture) of the zoning and development code that directly addressed the concerns raised by the Badseed Farm affair, providing definitions for what urban agriculture activity was allowed in residential, office/commercial, and manufacturing districts. Specifically, the ordinance now separates a Crop Agriculture use (formerly Agriculture, Nurseries and Truck Gardening) from three categories of Urban Agriculture: Home Garden, Community Garden, and Community Supported Agriculture. Crop agriculture and home and community gardens are permitted in all districts, while CSA farms with their attendant apprentices require a special use permit to locate in a residential district. (As a special use, the establishment of a CSA farm must be publicly announced, and neighbors within 300 feet of it can testify about it at a public hearing.) The ordinance also establishes Animal Agriculture as an agricultural use that is permitted, with restrictions, in all districts. (KCMO is lenient regarding livestock keeping, another legacy of its agricultural history. Chapter 14 of the Code of Ordinances allows a wide variety of livestock, regulating only the number of animals and their distance from neighboring buildings. Beekeeping has always been allowed.)

The amended zoning ordinance does not separate traditional community gardens from "market gardens," as has recently been done in Cleveland and Madison, Wisconsin. Instead, community gardens are considered collective endeavors in which selling what is grown is optional. Each of the three Urban Agriculture designations allows on-site sales; this important point reflects the city council's desire to make home gardens and community gardens sources for fresh food in low-income areas lacking quality retail outlets (KCCUA n.d.).

Neighboring municipalities are monitoring the impact of KCMO's actions before deciding whether to replicate them. KCMO officials hope that mandatory six- and 18-month ordinance reviews will showcase the positive aspects of the zoning changes. But other municipalities will also look to see if the regulations have effects on crime and property values in areas surrounding urban farms.

While the amended ordinance addresses the primary advocacy objectives of the urban agriculture community, some unresolved issues remain. Since advocates initially focused on less-controversial fruit and vegetable growing, they now plan to work on clarifying livestock-raising allowances. Local urban agriculture groups, planners, and city officials are justifiably proud of the expansion of the codes to include the new models of urban food production being used across the city.

Figure 4.6

Revenue-generating community-supported agriculture farms were noted as operating on the regulatory margins in Kansas City, Missouri, and governance problems arose because of a lack of definitions for urban agriculture types.

New Orleans: Grassroots Activism in Rebuilding the Food System

The catastrophe of Hurricane Katrina in late August 2005 severely damaged New Orleans's (pop. est. 354,850 post-Katrina) physical and social infrastructure. Over the course of five years, New Orleanians have overcome initial doubts over whether the city should even be rebuilt and dismissed the recommendation of planners that the future city occupy a smaller footprint.

In part because of the perceived failure of planners and other government agencies to take the lead in rebuilding New Orleans, many residents believe that the successes that the city has had in renewing itself have had less to do with the effectiveness of local government and more with the combined initiative of the many nonprofit organizations dedicated to one or more segments of the city's social and environmental structures. The current rebuilding of the city's food system—and the development of urban agriculture—is due primarily to grassroots creativity and action rather than government initiative. Planners in New Orleans are the followers, not the leaders, in re-creating the city's food system. This situation is in the process of changing, however, with the inauguration of Mitch Landrieu as mayor in May 2010 fostering new optimism.

The current rebuilding of New Orleans's food system—and the development of urban agriculture—is due primarily to grassroots creativity and action rather than government initiative.

New Orleans has a rich culinary history. The contributions of Native American, French Creole, Cajun, African, and Caribbean cultures, as well as modern Central American and Vietnamese influences, formed a distinctive array of cuisines based on sauces and spices mixed with produce that could be grown locally throughout the year. German immigrants farmed in the River Parishes upriver from New Orleans; Sicilian truck farmers from St. Bernard Parish grew Creole artichokes, tomatoes, and garlic. Coastal fisherman originally from the Canary Islands or China supplied oysters, shrimp, and crawfish (Sauder 1981). Citrus farms developed in Plaquemines Parish downriver, and strawberries were grown in Tangipahoa Parish across Lake Pontchartrain.

This local bounty was sold in a large number of urban markets—the French Market near Jackson Square began in the late 18th century and still operates today. Later municipal markets, such as St. Bernard and St. Roch, were the focus of residential neighborhoods elsewhere in the city; both were still active before Katrina. In addition, numerous curbside vendors sold produce out of small trucks.

The pre-Katrina food system in New Orleans thus combined neighborhood-scaled elements with conventional ones, such as superstores of 30,000 square feet or larger. In the rebuilding process, however, large retailers have been slow to return; by early 2008 there were only 18 full-service supermarkets unevenly distributed within the city—one per every 18,000 residents (New Orleans FPAC 2008).[7] Along with the loss of smaller neighborhood retailers, this created food deserts in heavily storm-damaged areas such as Broadmoor, Gentilly, the Seventh and Ninth Wards, and New Orleans East, where much of the city's black middle class lived. This situation has not significantly improved in the past two years.

Apart from the supermarket gap, however, the recovery of the New Orleans food system is being inspired by the national local food movement, and advocates such as the New Orleans Food and Farm Network (NOFFN; www.noffn.org) hope to increase the number of alternative food production and purchasing options beyond what existed pre-Katrina. Farmers markets, community gardens, and urban farms are under way or envisioned. And planners and other government representatives are the targets of ongoing food-system advocacy by individuals and nongovernmental entities.

Recent Urban Agriculture Developments. Just about every endeavor in New Orleans is affected by Hurricane Katrina's legacy, and efforts to improve urban agriculture are no exception. Before Katrina, Parkway Partners, a citywide greening organization, was the focal point of the local urban agriculture scene. As is typical for many community gardening organizations, Parkway Partners managed community gardens without many financial resources. In addition, the Vietnamese community in New Orleans East produced significant amounts of food in physical and social isolation from the rest of the city. By 2002, NOFFN—an organization dedicated to creating food policy, identifying gaps in food access, promoting urban agriculture, and supporting local producers—had been formed.

After the storm, NOFFN and other activists were quick to assess the state of the food system. Working outside of city government, they identified both short-term food-system actions for those returning to the city and long-term strategies to parallel the rebuilding of other systems. One important immediate action was the creation of digital neighborhood food maps, with updated locations of retailers, reopened restaurants, and revived and new community gardens. Some viewed urban agriculture as a necessary strategy in the immediate post-Katrina months since food was simply not otherwise available (Olopade 2009). The dramatic increase in blighted residential addresses and vacant parcels—from 19,000 parcels pre-Katrina to 43,755 as of November 2010—represented a significant opportunity for increasing urban food production (Plyer et al. 2010).

The widespread devastation of the food system, however, meant that all components, including urban agriculture, had to be addressed. Chief among food system advocates was the Food Policy Advisory Committee (FPAC), created by the New Orleans City Council in May 2007 to identify the barriers limiting the availability of healthy and nutritious food. FPAC submitted a report to the city council in March of the following year containing a series of recommendations to improve food access. As addressing the critical absence of supermarkets had become a priority, the report focused on food retailing; urban agriculture was only indirectly referenced in a recommendation to remove regulatory barriers to businesses selling fresh food (New Orleans FPAC 2008).

While supermarkets continue to be unevenly distributed, urban agriculture is slowly increasing its presence in the Crescent City's recovery (Olopade 2009; Bailkey 2009). Parkway Partners now manages close to 30 community gardens, an accomplishment considering that most of the pre-Katrina gardens were destroyed. The two-year-old Hollygrove Market and Farm—which is the product of a partnership between NOFFN and the

The recovery of the New Orleans food system is being inspired by the national local food movement, and advocates such as the New Orleans Food and Farm Network.

Carrollton-Hollygrove Community Development Corporation—supplies neighborhood residents with twice-weekly CSA market boxes and is the site of community garden plots and two urban farms. At a smaller scale, individuals have started small urban farms on vacant lots in the Central City and Mid-City neighborhoods, and a NOFFN-initiated backyard-garden program thrives across the Mississippi River in Algiers. And in the Uptown neighborhood at the Samuel J. Green Charter School, the Edible Schoolyard (http://edibleschoolyard.org) brings to New Orleans an organic school garden and kitchen classroom model developed by Alice Waters in Berkeley, California. In the Lower Ninth Ward, a neighborhood heavily devastated by Katrina's floodwaters, Our School at Blair Grocery and the Lowernine .org Garden Program practice urban agriculture within sight of block after block of vacant lots and abandoned homes.

In New Orleans East, the planned Viet Village Urban Farm project demonstrates the new engagement of the Vietnamese community with other New Orleanians to collectively advocate for urban farms. The project plan is a 2008 winner of an American Society of Landscape Architects award for excellence in analysis and planning, but it is on hold while its supporters seek permits for development of the site while also fighting a related battle over site contamination. (See Figure 4.7 and http://mqvncdc.org/page .php?id=18.)

Figure 4.7. *Viet Village Urban Farm site plan*

MQVN Community Development Corporation

The city's urban agriculture actors collaborate, but not as strongly as they might—and planners on the whole are framing their discussions of urban agriculture within the city's recently approved 20-year master plan. In addition, the raised profile of urban agriculture has brought in more players, and they compete for a limited amount of support dollars, primarily from private foundations. There is also some concern that certain urban farmers operating under the regulatory radar (e.g., those cultivating livestock) might, if discovered, make things difficult for those playing by the rules.

Urban Agriculture Advocacy. Community advocacy for urban agriculture immediately after Katrina took a backseat to efforts to restore an adequate level of food access across all New Orleans neighborhoods, primarily through the return of supermarkets and other retailers. The FPAC—composed of representatives from the retail sector, emergency food providers, local universities, and city agencies, among others—had few members focused primarily on promoting urban agriculture. In addition, the New Orleans urban agriculture community was less focused on seeking relevant policy changes in the immediate post-Katrina years and more interested in simply finding sites on which to grow and outlets through which to sell.

The FPAC's 2008 recommendations, mentioned above, only indirectly addressed urban agriculture by noting a zoning restriction on direct on-site sales from gardens and small urban farms in single-family residential districts (New Orleans 2008). The lack of clear language regarding on-site sales had long been a concern of the city's urban growers. On-site businesses selling produce are specifically prohibited if fruit and vegetable growing is considered an accessory use, but where agriculture is a primary use, the code is silent. Clarification on this issue is needed; removing the restriction could facilitate residents' access to fresh food in neighborhoods lacking adequate retail outlets.

The major advocacy challenge for New Orleans city farmers today is bringing urban agriculture to the attention of city officials in the face of other pressing social needs.

The major advocacy challenge for New Orleans city farmers today is bringing urban agriculture to the attention of city officials in the face of other pressing social needs—providing affordable housing, bringing down the murder rate, finding enough money for the city's recreation department—all in the midst of severe budget constraints. While this will be no easy task, advocates still see opportunities in the city's professed desire to chart a post-Katrina path of sustainable rebuilding and reuse of the city's vacant parcels.

Urban Agriculture and Local Planning Mechanisms. Changes in public planning mechanisms in the city of New Orleans currently have the attention of the urban agriculture community. Specific focus areas include: (1) Plan for the 21st Century: New Orleans 2030, the new 20-year master plan approved by the city council in August 2010; (2) the accompanying revamping of the city's comprehensive zoning ordinance; and (3) gaining access to vacant land through the New Orleans Redevelopment Agency (NORA).

Kimberley Hodgson

Master Plan

In 2008, New Orleans voters approved a binding referendum amending the city charter to give the forthcoming master plan, unlike previous comprehensive plans, the force of law—meaning that all future land-use and zoning actions must conform to the plan's goals and objectives. Because urban agriculture is well represented in the plan, supporters hope urban farms will become prevalent across the city.

The master plan, developed by the planning commission through an extensive public participation process with assistance from Boston-based consultants Goody Clancy, is extensive and visionary, with a strong emphasis on sustainable growth and development. Urban agriculture appears in several sections. The 2008 FPAC food-access recommendations are echoed in Chapter 8, Health and Human Services, in the goal "Access to fresh, healthy food for all residents." In Chapter 14, Land Use, allowing "urban agriculture in appropriate locations" is a recommended action to "promote smart growth land use patterns in New Orleans and the region." Finally, the plan specifies a number of policies and action items in Chapter 13, Environmental Quality, where urban agriculture is one of eight major action areas (Goody Clancy 2010). While the new plan enables many opportunities for urban agriculture, challenges remain in implementing the action recommendations and creating mechanisms to secure long-term tenure for urban farm sites.

Comprehensive Zoning Ordinance

The development of a new comprehensive zoning ordinance (CZO) is the next step in the process begun with the development of the new master plan. The current New Orleans CZO, as in many cities, is an assemblage of ad hoc amendments, changes, and rewrites collected over decades without the framework of an overarching comprehensive plan. Food-system and urban-agriculture supporters are prepared to work with the planning commission and its zoning code consultants to develop language consistent with the appropriate master-plan goals; for example, addressing the lack of clarity about selling from urban farms in residential districts, as described earlier, and adding provisions for livestock keeping, which is currently regulated through the city's health code.

A goal of the New Orleans Redevelopment Authority is to make blighted and adjudicated properties available for urban farms.

Vacant Land

As in other cities with significant vacant-land inventories, the New Orleans urban agriculture community sees the current amount of open land as an opportunity. The city's priority is to return vacant properties to residential or commercial use, but in the hardest-hit areas of the Lower Ninth Ward, Gentilly, and elsewhere, supply exceeds current demand, providing prospects for alternative uses, such as urban agriculture.

The agency responsible for vacant and adjudicated property is the New Orleans Redevelopment Authority (NORA), working through a cooperative agreement with the city council. Currently, NORA owns 4,800 cleared-title properties acquired from the state program known as Road Home. These properties have completed the complicated review

Kimberley Hodgson

process through which the legal status of the land is established, and can now be made more easily available to the public. NORA can also file expropriation lawsuits to acquire parcels designated by officials as "blighted," another complicated process. NORA currently manages two land-distribution programs: one, the city council–created Lot Next Door program, gives those owning property adjacent to NORA-owned parcels first opportunity to acquire them, provided they pay the appraised fair-market value and plan to own the property for at least five years. Under the second program, NORA will issue a request for proposals (RFP) to citizens interested in acquiring NORA-owned parcels in particular neighborhoods. An applicant must state objectives for use of the property, funding sources, and more. If, after review, NORA and neighborhood representatives accept the proposal, the applicant can purchase the property for its appraised value.

Currently, these two mechanisms can be used to acquire land for urban agriculture, but they operate largely on a site-by-site basis, and housing remains a priority. However, a larger framework for dispensing land parcels for urban agriculture could be provided by the new master plan, which identifies NORA as the sole or partner agent to fulfill three plan recommendations: to create an inventory of possible and suitable garden or farm sites; to make blighted and adjudicated property available for urban farms; and to explore community orchards as an interim land use.

Urban Agriculture and Local Government. New Orleans city government has not historically been a champion of urban agriculture, although in recent years several city-council members have become supporters. Generally, city government is neither an ally nor an obstacle to the expansion of urban agriculture, but advocates note much room for improvement—specifically better communication with and access to city officials. Nevertheless, optimism has been growing since the inauguration of Mitch Landrieu. Following Landrieu's election in February 2010, his transition team convened 17 task forces of community leaders and experts on critical urban issues and charged them with recommending quick-impact actions for the administration's first 100 days. Food system and urban agriculture advocates were members of the Sustainable Energy and Environmental Task Force, whose report recommended the implementation of urban food gardens, community markets and fresh food retail initiatives, and other recovery projects already approved by the city council, funded by a small portion of the $411 million in federal recovery Community Development Block Grant (CDBG) funds awarded to the city after it completed a comprehensive recovery plan in 2007. The task force also recommended that the administration address the bureaucratic obstacles stalling the Viet Village Urban Farm.

Summary. In summary, urban agriculture is positioned to establish a niche role in the overall recovery of New Orleans. However, much of what has been described here—the new master plan and the revamped CZO, the possibility for support from the Landrieu administration, the available but unreleased federal CDBG funding—signifies potential rather than current, measurable impact. In the meantime, individual on-the-ground projects and the community-based grassroots activism driving neighborhood recovery will maintain the momentum for urban agriculture until larger, citywide planning mechanisms can direct government support to it.

CONCLUSION

The case-study communities presented here offer rich and varied examples of how public-sector planners, agency staff, local policy makers, nongovernmental organizations, and community-based groups are actively collaborating and creating innovative strategies to support and enhance urban agriculture as an important component of broader community planning issues: urban resilience, sustainability, redevelopment, brownfields remediation and reuse, health and wellness, and capacity building and empowerment. Chapter 5 provides a summary and brief discussion of key lessons learned from these communities.

ENDNOTES

1. Dr. Mary Hendrickson and Dr. Robert D. Heffernan of the University of Missouri–Columbia's Food Circles Networking Project have documented the impacts of the increasing vertical integration and consolidation of the industrial food system over more than a decade. Their research may be accessed at www.foodcircles.missouri.edu/consol.htm.

2. Community Food Animators promote best practices that help bring to life new projects and food services, such as community kitchens, community gardens, fresh food markets, and enhancement to the emergency food sector. See www.foodshare.net/animators01 .htm.

3. Phase I assessments—also known as "All Appropriate Inquiries"—include investigations of historical land-use patterns and ownership, local geology and hydrology, and current land uses of a property. Phase I assessments do not involve any subsurface investigations of a site. Phase II assessments investigate the degree and extent of surface and subsurface contamination of a site to determine the degree of environmental impairment. They can include soil and groundwater investigations for suspected contaminants based on the Phase I study. Projections of remediation costs are also sometimes made.

4. Recent U.S. Department of Health and Human Services estimates show that more than 66 percent of adults are overweight or obese and that 17 percent of adolescents and 19 percent of children are overweight.

5. See www.seattle.gov/council/conlin/food_initiative.htm. The resolution is number 31019, adopted April 2008; available at http://clerk.ci.seattle.wa.us/~archives/Resolutions/Resn_31019.pdf.

6. Transition Towns originated in the United Kingdom in 2005, as a "community-led response to climate change, fossil fuel depletion and increasingly, economic contraction," but the movement has since spread worldwide. See www.transitionnetwork.org/support/what-transition-initiative and www.villagevancouver.ca/group/villagevancouverfoodworkinggroup.

7. For comparison, the national average ratio is one supermarket for every 8,800 residents. The pre-Katrina ratio in New Orleans was one supermarket per 12,000 residents (Goody Clancy 2010).

CHAPTER 5

Planning for Urban Agriculture:
Lessons Learned

 The case studies in this report draw on extensive research and interviews with representatives of local government agencies, universities, and community-based and nonprofit organizations in 11 North American cities. This chapter identifies lessons learned: common approaches to the development and implementation of municipal studies and assessments, plans, policies, and programs to support and encourage all types of agriculture within urban environments.

Although no one prescription exists for integrating agriculture into the urban fabric through planning practice, the lessons here point to critical aspects of the process. The steps that planners and local governments actually take will depend upon their particular local and regional contexts, as well as their knowledge and understanding of urban agriculture and the food system.

(1) Urban agriculture can positively contribute to a healthy, resilient community, especially when combined with other planning strategies.

Though urban agriculture alone cannot solve all of a community's problems, it is an important complement to other strategies and objectives. Planners should consider how urban agriculture can be used to support environmental, community-development, and quality-of-life goals. For example, community gardens can be incorporated into affordable-housing developments to provide social spaces, foster tenant interaction, and create opportunities for physical activity and healthy eating. Commercial urban farming operations—whether sited on underutilized properties in built-out areas or on vacant properties in cities—can reduce fertilizer and pesticide use if replacing lawns; mitigate stormwater runoff if replacing pavement; minimize energy use from mowing and other land-maintenance activities; and provide much-needed skilled and unskilled urban job opportunities. Rooftop vegetable and fruit gardens, as well as rooftop beekeeping, contribute to food security and public health as well as urban cooling, greening, and stormwater management. School gardens can help fight obesity by educating children about healthy eating. The case studies in this report highlight how cities are using a range of urban agriculture activities to realize these outcomes.

SOLEfood

(2) Public interest, support, and engagement create a foundation for successful urban-agriculture planning, policy development, and implementation.

In some case-study communities, such as Cleveland, Vancouver, and Philadelphia, public pressure was instrumental in prompting local government action to support and expand urban agriculture through planning, programmatic, or regulatory activities. At times, a robust public-engagement process facilitated by local government (e.g., Minneapolis, Kansas City, and Seattle) or by food policy or urban agriculture coalitions (e.g., Cleveland, Vancouver, and Philadelphia) brought urban agriculture to the forefront of public policy discussions. But to many local governments, urban agriculture remains a novel activity that contrasts with conventional understandings of urban land use and food supply chains. And as with other innovations, a certain degree of public-sector skepticism must be overcome before this use becomes an accepted piece of the urban fabric. As evidenced by several of the case-study communities, growing popular interest in and mainstream media attention to urban agriculture is helping to accelerate a shift in public-sector thinking.

(3) Engaged political leadership and support are important to the development and implementation of urban agriculture policies and programs.

Politically, urban agriculture benefits greatly from champions advancing its cause. The support of local policy makers—such as mayors, city or county council members, and planning commissioners—can jumpstart local government action and streamline the policy-making process. (The importance of local actors simply wanting to match the innovative practices they see in other cities cannot be overlooked.) In several of the case-study communities, political leaders issued municipal mandates, motions, or challenges that (a) sanctioned the efforts of local government staff, coalitions, or community-based organizations already engaging in programmatic and regulatory change to support urban agriculture, and (b) prompted or required other local government staff to participate in this change. By publicly highlighting connections between urban agriculture and sense of community, health, environmental quality, and other issues, policy makers can draw attention to and enhance residents' understanding of urban agriculture.

(4) Urban agriculture coalitions play central roles in organizing a diverse range of stakeholders and effectively communicating their needs and concerns to local government staff and policy makers.

Food policy councils and local urban-agriculture coalitions play important roles in coordinating efforts, aligning goals, facilitating communication among diverse stakeholders and local government, and advocating for policy and program changes. They often have strong and direct connections to community groups and residents, particularly underserved and marginalized populations, and are aware of specific health, nutrition, economic, education, and other social needs. These groups offer both informal and formal mechanisms, such as social events and structured meetings, for communication and collaboration, creating and maintaining spaces in which practitioners can exchange ideas about specific projects and forums for policy discussions about what local governments can do to support urban agriculture. Many of the case-study communities have some type of urban agriculture–focused coalition, whether a subcommittee of a food policy council (or similar entity) or a stand-alone network of urban agriculture stakeholders organized by nonprofit organizations whose own missions span those of individual organizations. Coalitions can identify stakeholder needs and concerns and communicate them effectively to local government staff and policy makers.

(5) Local government committees are necessary to encourage and facilitate cross-departmental communication and coordination regarding urban agriculture initiatives.

Because urban agriculture cuts across the responsibilities of multiple local government sectors—planning, public health, economic and community development, water, solid waste, brownfields, transportation, and others—an umbrella committee representing multiple departments can increase intragovernmental communication, coordination, and collaboration related to urban agriculture. Such committees allow individual departments to address specific issues related to urban agriculture, while understanding how individual programs, projects, and policies can support urban agriculture and how urban agriculture can be integrated into individual departments' activities. A committee of this type also encourages the exchange of practical information and the sharing of ideas, offering resolutions to any issues, tensions, or conflicts that arise. In particular, intragovernmental committees mandated by elected officials can facilitate coordination across governmental departments by tasking each department with specific actions within a broader urban agriculture framework. These could include assessing regulatory barriers to urban agriculture or revising internal staff policies that prevent a local government from supporting urban agriculture. Newly created offices or departments of sustainability, for example, are positioned to incorporate urban agriculture into broader municipal strategies because of its overlap with social, economic, and environmental sustainability, as evidenced in Philadelphia, Vancouver, Toronto, and Cleveland.

(6) Urban agriculture proliferates in communities with a wide range of policies and programs to support the diversity of urban agriculture types, sizes, and scales and its integration into the urban fabric.

Absent a diverse range of policies and programs supporting urban agriculture in a variety of types, sizes, and scales, planners may not have sufficient capacities to incorporate urban agriculture into the urban fabric. Conversely, when policy makers and local government staff simply equate urban agriculture with community gardens or, in other cases, large-scale commercial farming operations, they miss urban agriculture's complexity and diversity and may not embrace the policies and programs necessary to support it in different community contexts.

With all the benefits and opportunities urban agriculture presents, many people in the case-study communities are beginning to consider it on a par with traditional amenities such as housing, parks and open space, and infrastructure. They value urban agriculture's ability to adapt to different contexts. For example, large vacant parcels may be well suited to commercial urban farms, underutilized institutional properties may lend themselves to educational gardens or hybrid urban farms, and existing parks in urban or suburban neighborhoods may have space for community gardens. Because urban agriculture is compatible with many existing land uses and can be creatively and innovatively molded to fit different contexts, it can provide many kinds of residents with an important public amenity.

Urban agriculture–friendly zoning regulations that allow gardening and farming, farmers markets and on-site produce sales, and animal keeping and composting in all appropriate zoning districts are important, but policies and programs that go beyond land use are also required for the success of these activities. As noted, many local government agencies and departments have roles to play in creating supportive regulatory environments and delivering programs to urban agriculture practitioners and the general public.

(7) Public planning for urban agriculture does not require a special skill set; traditional planning tools and approaches can facilitate its implementation.

This report asserts that urban agriculture, like any other land use, can be guided and regulated through comprehensive planning and land-use practices. Though knowledge and policy gaps currently exist on how best to regulate certain practices such as composting and animal keeping, planners can learn how to develop appropriate land-use controls by emulating pioneering ordinances of other cities. Community and neighborhood planners can introduce the dimensions and benefits of urban agriculture in community visioning sessions and then facilitate practical discussions of integrating implementation strategies into community goals.

(8) Food-system assessments and land inventories help justify the need to plan for urban agriculture.

Most of the case-study communities initiated assessments of their local food systems and by doing so identified urban agriculture as a central issue warranting further exploration. In some cases, the assessments were conducted by local governments; however, in other cases where funding or governmental interest was scarce, they were conducted by a third party—typically a local nonprofit organization or university. These assessments provided justifications for subsequent plan making and policy development, particularly in communities where local government staff or policy makers were not initially receptive to urban agriculture. When strong public and local nonprofit support for urban agriculture surfaces in comprehensive, sustainability, or neighborhood planning processes, this can prompt policy makers and local government staff to investigate regulatory barriers that stand in the way of practitioners.

(9) Land values often dictate local government policy and programmatic approaches to urban agriculture.

Case-study communities in both high and low land-value areas (e.g., Seattle and Cleveland, respectively) are turning to market gardens, commercial urban farms, and hybrid urban-agriculture programs as ways to simultaneously manage stormwater, reuse vacant or underutilized lots, provide access to open space, and promote green-jobs training. However, both high and low land costs pose specific challenges.

High Cost. The most significant barriers to expanding urban agriculture in dense or built-out cities are high land costs and intense development pressures. Because these cities have little vacant land and relatively few brownfield sites, governments planning for urban agriculture need to creatively integrate it within the existing environment. In this context, both public and privately owned lands—such as parks, municipal government property, schools, or hospital grounds—represent potential farm or garden sites. Urban agriculture can also be incorporated within private development projects and nontraditional spaces, such as rooftops, to create a diversity of urban agriculture types within densely populated neighborhoods. Promoting and integrating food production within public green-roof mandates and incentives is a promising approach that is already taking hold in Milwaukee and Chicago.

Low Cost. Often, the most significant barriers to expanding urban agriculture beyond community gardens in cities with an abundance of vacant property are land acquisition and disposition policies oriented toward physi-

Kimberley Hodgson

cal redevelopment, as well as actual or perceived soil contamination. Absentee landowners can make it difficult for a city to acquire land for urban agriculture and transfer it to nonprofit or other ownership. Local governments may view vacant land only through the lens of future development potential; planners can use vacant land inventories to identify areas where future higher revenue–generating land uses are best located and areas where urban agriculture could provide significant community benefit. Soil contamination can often be difficult to assess and costly to clean up. Case-study communities identified needs for more systematic identification and evaluation of land suitable for urban agriculture, communitywide brownfield assessments, and specific standards and guidelines for the proper remediation of brownfields for urban agricultural use (or, in cases where this is not feasible, standards for effective raised-bed or container growing systems).

(10) Successful urban-agriculture policies are often part of broader community food-system agendas.

Creating urban environments that are favorable and supportive of all forms of urban agriculture requires planners to consider how urban agriculture fits into the larger local and regional food system. This represents a broader level of integrated understanding: how to fit urban agriculture within a food supply and consumption system that has long been taken for granted and how to fit the food system into a larger social infrastructure. Case-study communities that have successfully removed regulatory barriers for urban agriculture tend to have strong foundations in community food-systems planning, including traditions of food-producing community gardens or newer efforts such as food policy councils. Urban agriculture thus becomes one part of a comprehensive approach to integrating food system considerations into municipal policy and decision-making processes.

CONCLUSION

These are promising times for urban agriculture in North America. Each recent growing season has brought an abundance of new projects, and the increasing recognition of urban agriculture in the popular media seems to indicate its legitimization. Yet popular awareness does not by itself lead to public policy change. While urban agriculture is not foreign to planning history, thought, or practice, planners today too often lack a sufficient and functional understanding of urban agriculture.

Urban agriculture is part of a larger, growing movement with the potential to influence the food-related choices of all North Americans, rich and poor.

In both the United States and Canada, planning for urban agriculture is developing in an organic fashion, with certain cities and regions more advanced than others. There are few accurate predictors for why some cities and regions are urban agriculture leaders; a progressive social environment can be a factor, but so too can the existence of critical amounts of underutilized land. But urban agriculture appears to be well on the way to becoming an accepted part of the urban fabric throughout North America. The case-study communities featured here should help planners develop practical strategies and approaches that can contribute to the steadily growing body of practical knowledge, fueling the creation of a paradigm for urban agriculture planning.

Planners must realize that enabling urban agriculture is more than a response to citizens' demands for opportunities to grow food in closer proximity to their tables. Urban agriculture is part of a larger, growing movement with the potential to influence the food-related choices of all North Americans, rich and poor. Planners now have ample opportunities and reasons to influence the future of the food system. Urban agriculture represents an important opportunity to grow healthier, more sustainable, and more resilient communities.

Growing Home's Wood Street Urban Farm, Chicago

Appendix 1. Urban Agriculture Components in Food Charters

Jurisdiction	Plan Name, Date, and Link	Philadelphia	Prince Albert, Sask.	Toronto, Ont.	Vancouver, B.C.

Food System Topics

Topic	Philadelphia	Prince Albert	Toronto	Vancouver
community engagement	x	x	x	x
food waste and disposal				x
food literacy and education	x	x	x	
food access and availability	x	x	x	x
food retail	x	x		x
food distribution	x	x		x
food processing	x	x		x
rural agriculture		x		
other topics			x	x
health/nutrition education	x	x	x	x
community development	x		x	x
environmental stewardship			x	
agricultural skills and knowledge				
agricultural practices			x	
financial assistance				
water	x			
land tenure				
uncontaminated soil				
growing space	x	x	x	
farm animals				
chickens				
bees				
orchards				
rooftop urban agriculture				
commercial farms	x			
commercial gardens				
edible landscaping				
private gardens	x			
institutional gardens	x			
community gardens	x	x	x	

Urban Agriculture Topics Addressed by the Plan / **How does plan support urban agriculture?**

Jurisdiction	Plan Name, Date, and Link	How does plan support urban agriculture?
Philadelphia	Philadelphia Food Charter, 2008 — www.leadershipforhealthycommunities.org/images/stories/philadelphia_food_charter1.pdf	Highlights urban agriculture as supportive measure toward achieving citywide sustainability initiative. City is committed to developing and coordinating municipal food and urban agriculture policy, and it encourages community gardening and farming, health and nutrition education, and institutional and private gardens, all initiatives supporting urban agriculture
Prince Albert, Sask.	A Food Charter, 2003 — www.foodsecuritynews.com/presentations/PrinceAlbert_Food_Charter.pdf	Encourages urban agriculture through community gardens, which can connect the rural food producer and the urban consumer
Toronto, Ont.	Food Charter, 2001 — www.toronto.ca/food_hunger/pdf/food_charter.pdf	Highlights food access and security issues, food purchasing and nutrition programs, and protecting vital agricultural land as ways of encouraging and supporting urban agriculture
Vancouver, B.C.	Vancouver Food Charter, 2007 — http://vancouver.ca/commsvcs/socialplanning/initiatives/foodpolicy/policy/charter.htm	Outlines the need for a municipal food policy that benefits and supports the economic, ecological, and social well-being of the city and its residents. Promotes expanding urban agriculture and food recovery opportunities

Appendix 2. Urban Agriculture Components in Local Comprehensive Plans

Jurisdiction	Plan Name, Date, and Link	Food System Topics									Urban Agriculture Topics Addressed by the Plan																				
		community engagement	food waste and disposal	food literacy and education	food access and availability	food retail	food distribution	food processing	rural agriculture	other topics	health/nutrition education	community development	environmental stewardship	agricultural skills and knowledge	agricultural practices	financial assistance	water	land tenure	uncontaminated soil	growing space	farm animals	chickens	bees	orchards	rooftop urban agriculture	commercial farms	commercial gardens	edible landscaping	private gardens	institutional gardens	community gardens
Asheville, N.C.	Asheville City Development Plan 2025, 2003. www.ashevillenc.gov/business/development_services/planning_zoning/default.aspx?id=1144	×		×	×		×	×		×	×	×	×	×	×			×											×	×	×
Bar Harbor, Me.	Town of Bar Harbor Comprehensive Plan, 2007. www.barharbormaine.gov/compplan			×	×					×						×		×	×							×					
Berkeley, Calif.	City of Berkeley General Plan: A Guide for Public Decision-Making, 2001. www.ci.berkeley.ca.us/contentdisplay.aspx?id=494	×		×	×		×				×	×								×									×	×	×
Boise, Idaho	Blueprint Boise, 2010. https://cityofboise.org/BluePrintBoise	×								×			×	×			×													×	×
Dane County, Wis.	Comprehensive Plan, 2007. http://danedocs.countyofdane.com/webdocs/PDF/PlanDev/ComprehensivePlan/CH5_Agriculture.pdf	×	×	×	×	×	×	×	×			×	×	×	×																
Dillingham, Alaska	City of Dillingham Comprehensive Plan Update and Waterfront Plan, 2010. www.agnewbeck.com/pdf/bristolbay/Dillingham_Comp_Plan/DLG_COMPlan_FINAL_email.pdf	×	×	×	×		×	×		×	×	×		×	×				×	×							×		×	×	×
Fitchburg, Wis.	City of Fitchburg Comprehensive Plan, 2009. www.city.fitchburg.wi.us/departments/cityHall/planning/Comprehensive.php	×		×	×				×			×								×											×
Grand Rapids, Mich.	City of Grand Rapids Master Plan, 2002. www.grand-rapids.mi.us/index.pl?page_id=632	×			×					×		×	×		×					×									×	×	×
Greendale, Wis.	Village of Greendale Comprehensive Plan, 2009. www.greendale.org/comprehensive_plan.htm		×		×		×													×					×					×	×

(continued)

Jurisdiction	Plan document
King County, Wash.	King County Comprehensive Plan, 2008 www.kingcounty.gov/property/permits/codes/growth/CompPlan.aspx
Madison, Wis.	City of Madison Comprehensive Plan, 2007 www.cityofmadison.com/planning/comp
Marin County, Calif.	Marin Countywide Plan, 2007 www.co.marin.ca.us/depts/CD/main/fm/cwpdocs/CWP_CD2.pdf
New Orleans	Plan for the 21st Century: New Orleans 2030, 2010 www.nolamasterplan.org/documentsandresources.asp#C11
Richmond, Calif.	City of Richmond General Plan Update (draft), 2009 www.cityofrichmondgeneralplan.org/Content/10020/preview.html
San Francisco	San Francisco General Plan, 2009 www.sf-planning.org/ftp/General_Plan/index.htm
Santa Rosa, Calif.	Santa Rosa General Plan 2035, 2009 http://ci.santa-rosa.ca.us/doclib/Documents/2035_General_Plan.pdf
Tacoma, Wash.	Comprehensive Plan, 2010 (proposed amendment) http://cms.cityoftacoma.org/Planning/2010%20Annual%20Amendment/PublicHearing/C2b_UrbanForest.pdf
Washtenaw County, Mich.	A Comprehensive Plan for Washtenaw County: A Sense of Place, A Sustainable Future, 2004 www.ewashtenaw.org/government/departments/planning_environment/comp_plan/frontpage
Watsonville, Calif.	Watsonville VISTA 2030, 2005 www.ci.watsonville.ca.us/departments/cdd/general_plan/gp/0100%20dvrs_pop.pdf
Wayne County, N.C.	Wayne County Comprehensive Plan, 2009 www.waynegov.com/16581031616432657/lib/16581031616432657/ReAdoptedPlan.doc
Yuma, Ariz.	City of Yuma 2002 General Plan, 2002 www.yumaaz.gov/Documents/COY_GeneralPlan2002.pdf

(continued)

113

Appendix 2. Urban Agriculture Components in Local Comprehensive Plans

Jurisdiction	Plan Name, Date, and Link	How does plan support urban agriculture?
Asheville, N.C.	Asheville City Development Plan 2025, 2003 www.ashevillenc.gov/business/development_services/planning_zoning/default.aspx?id=1144	Supports creating urban parks and gardens, greenways, and riverways
Bar Harbor, Me.	Town of Bar Harbor Comprehensive Plan, 2007 www.barharbormaine.gov/compplan	Describes how agriculture preservation was historically dedicated to providing locally grown produce and protecting natural resources, including prime agricultural soils. One of the greatest opportunities for sustaining agricultural resources is by increasing agricultural activity among small-scale local farms.
Berkeley, Calif.	City of Berkeley General Plan: A Guide for Public Decision-Making , 2001 www.ci.berkeley.ca.us/contentdisplay.aspx?id=494	Open space and recreation element addresses growing need to include a broader range of community-based amenities and rising demand for additional space for community gardens, private gardens, and farmers markets. The plan gives high priority to maintaining public open spaces for community gardens and encouraging community groups to organize, design, and manage community gardens, particularly in spaces not suitable for housing, parks, pathways, or recreational facilities.
Boise, Idaho	Blueprint Boise, 2010 https://cityofboise.org/BluePrintBoise	Through land use, economic development, natural resources planning, and growth management, the plan highlights measures to preserve opportunities for urban agriculture, including community engagement.
Dane County, Wis.	Comprehensive Plan, 2007 http://danedocs.countyofdane.com/webdocs/PDF/PlanDev/ComprehensivePlan/CH5_Agriculture.pdf	Agricultural, natural, and cultural resources chapter includes strong farmland protection and local food goals and policies
Dillingham, Alaska	City of Dillingham Comprehensive Plan Update and Waterfront Plan, 2010 www.agnewbeck.com/pdf/bristolbay/Dillingham_Comp_Plan/DLG_COMPlan_FINAL_email.pdf	Supports local food production, processing, and distribution.
Fitchburg, Wis.	City of Fitchburg Comprehensive Plan, 2009 www.city.fitchburg.wi.us/departments/cityHall/planning/Comprehensive.php	Encourages urban agriculture through community gardens and farmers markets
Grand Rapids, Mich.	City of Grand Rapids Master Plan, 2002 www.grand-rapids.mi.us/index.pl?page_id=632	Encourages reusing vacant properties as community gardens. Parks and open space can provide urbanites with a greater appreciation for and understanding of natural systems and environmental stewardship. Encourages public-private partnerships with neighborhood organizations and civic group to initiate open space improvements
Greendale, Wis.	Village of Greendale Comprehensive Plan, 2009 www.greendale.org/comprehensive_plan.htm	Plan supports urban agriculture through community gardens, food waste and disposal, backyard gardens, rooftop gardens, and schoolyard greenhouses

(continued)

Location	Plan / Source	Description
King County, Wash.	King County Comprehensive Plan, 2008 — www.kingcounty.gov/property/permits/codes/growth/CompPlan.aspx	Explores ways of creating and supporting community gardens, farmers markets, produce stands, and other similar community-based food growing projects to provide and improve access to healthy food for all rural residents
Madison, Wis.	City of Madison Comprehensive Plan, 2007 — www.cityofmadison.com/planning/comp	Emphasizes urban agriculture through community gardens and private gardens and identifies the potential roles of community gardens in neighborhood improvement efforts and steps the city can take to facilitate these efforts
Marin County, Calif.	Marin Countywide Plan, 2007 — www.co.marin.ca.us/depts/CD/main/fm/cwpdocs/CWP_CD2.pdf	Working landscapes that produce food and other agricultural products are important to Marin County's history. Edible landscaping, community gardening, and the promotion of community education are some of the policies recommended to sustain urban agriculture in the county.
New Orleans	Plan for the 21st Century: New Orleans 2030, 2010 — www.nolamasterplan.org/documentsandresources.asp#C11	Plan encourages promoting the development of community gardens on both private and public land, creating opportunities to incorporate environmentally valuable permanent and temporary use of vacant land, and design and promote other urban green spaces. Plan's findings conclude that the desire for urban agriculture is steadily growing, but regulatory barriers remain challenging.
Richmond, Calif.	City of Richmond General Plan Update (draft), 2009 — www.cityofrichmondgeneralplan.org/Content/10020/preview.html	Community health and wellness element defines goals for promoting healthy lifestyles, e.g., access to community gardens; identifies a broad spectrum of community assets, e.g., urban produce stands and farmers markets, that promote healthy food choices; and further defines the connections between urban agriculture and food security, social cohesion, sustainability, and economic development
San Francisco	San Francisco General Plan, 2009 — www.sf-planning.org/ftp/General_Plan/index.htm	Recreation and open space element highlights the importance of prioritizing development of and access to urban community gardens
Santa Rosa, Calif.	Santa Rosa General Plan 2035, 2009 — http://ci.santa-rosa.ca.us/doclib/Documents/2035_General_Plan.pdf	Encourages using city- or county-owned land for neighborhood and community parks. Community gardens could be integrated to these areas.
Tacoma, Wash.	Comprehensive Plan, 2010 (proposed amendment) — http://cms.cityoftacoma.org/Planning/2010%20Annual%20Amendment/PublicHearing/C2b_UrbanForest.pdf	Proposed comp plan amendment for urban forestry includes urban agriculture
Washtenaw County, Mich.	A Comprehensive Plan for Washtenaw County: A Sense of Place, A Sustainable Future, 2004 — www.ewashtenaw.org/government/departments/planning_environment/comp_plan/frontpage	Encourages agriculture preservation and local food production
Watsonville, Calif.	Watsonville VISTA 2030, 2005 — www.ci.watsonville.ca.us/departments/cdd/general_plan/gp/0100%20dvrs_pop.pdf	Section 10, A Diverse Population, communicates the need to provide residents with access to safe and healthy food by supporting farmers markets and encouraging the development of community gardens on surplus land throughout the city.
Wayne County, N.C.	Wayne County Comprehensive Plan, 2009 — www.waynegov.com/165810316164326657/lib/165810316164326657/ReAdoptedPlan.doc	Encourages agriculture preservation and centralized water services and reuse technologies to support appropriate forms of agriculture; supports continued growth and prosperity of agribusiness throughout the county
Yuma, Ariz.	City of Yuma 2002 General Plan, 2002 — www.yumaaz.gov/Documents/COY_GeneralPlan2002.pdf	Supports local food production and community gardens

Appendix 3. Urban Agriculture Components in Local Sustainability Plans

Jurisdictions and plans:
- **San Francisco** — Sustainable San Francisco, 1997 — www.sustainable-city.org
- **Baltimore** — The Baltimore Sustainability Plan, 2009 — www.baltimorecity.gov/LinkClick.aspx?fileticket=DtRcjL%2flBcE%3d&tabid=128
- **Kansas City, Mo.** — City of Kansas City Climate Protection Plan, 2008 — www.marc.org/Environment/airQ/pdf/CP-Plan-7-16-08.pdf
- **Philadelphia** — Greenworks Philadelphia, 2009 — www.phila.gov/green/greenworks/PDFs/GreenworksPlan002.pdf
- **Cleveland** — Reimagining a More Sustainable Cleveland, 2008 — www.cudc.kent.edu/shrink/Images/reimagining_final_screen-res.pdf
- **Philadelphia** — Green Plan Philadelphia (forthcoming) — www.greenplanphiladelphia.com
- **Alexandria, Va.** — Eco-City Alexandria Environmental — http://ecocity.ncr.vt.edu/actionplan.html

Topic	San Francisco	Baltimore	Kansas City, Mo.	Philadelphia (Greenworks)	Cleveland	Philadelphia (Green Plan)	Alexandria, Va.
Food System Topics							
community engagement	×	×		×			
food waste and disposal		×	×		×		×
food literacy and education		×	×				
food access and availability	×	×	×	×	×		×
food retail	×	×	×				
food distribution	×	×	×	×	×		
food processing	×	×	×	×	×		
rural agriculture		×			×		
other topics			×	×			×
Urban Agriculture Topics Addressed by the Plan							
health/nutrition education	×	×		×			
community development	×	×		×			
environmental stewardship		×	×		×		
agricultural skills and knowledge	×	×		×	×		
agricultural practices		×		×	×		
financial assistance		×	×	×			
water				×	×		×
land tenure					×		
uncontaminated soil							
growing space	×	×		×			
farm animals					×		
chickens					×		
bees							
orchards							
rooftop urban agriculture	×			×			
commercial farms				×	×		
commercial gardens							
edible landscaping		×					
private gardens	×	×		×			
institutional gardens	×	×		×			
community gardens	×	×		×	×		

(continued)

Appendix 3. Urban Agriculture Components in Local Sustainability Plans

Jurisdiction	Plan Name, Date, and Link	How does plan support urban agriculture?
San Francisco	Sustainable San Francisco, 1997 www.sustainable-city.org	Promotes urban agriculture through community and private gardens, adequate growing spaces, and food access issues
Baltimore	The Baltimore Sustainability Plan, 2009 www.baltimorecity.gov/LinkClick.aspx?fileticket=DtRcjL%2flBcE%3d&tabid=128	Supports developing an urban agriculture plan that will promote healthy, local, and, where possible, organic food production. The plan should identify the predicted demand for urban farmed food and recommend location and distribution of urban agricultural institutions.
Kansas City, Mo.	City of Kansas City Climate Protection Plan, 2008 www.marc.org/Environment/airQ/pdf/CP-Plan-7-16-08.pdf	Promotes residential neighborhood and metropolitan food production, expansion of the existing urban forestry program, and matching participation from the private sector
Philadelphia	Greenworks Philadelphia, 2009 www.phila.gov/green/ greenworks/PDFs/ GreenworksPlan002.pdf	Promotes access to fresh, healthy foods, innovative financing for redeveloping open spaces, expanding the number of neighborhood farmers markets, fostering commercial farming, and creating an urban agriculture workforce strategy to grow green jobs
Cleveland	Reimagining a More Sustainable Cleveland, 2008 www.cudc.kent.edu/shrink/ Images/reimagining_final_screen-res.pdf	Supports urban agriculture and commodity farming, establishing water rates that promote agricultural uses and improving community composting. Encourages developing urban agriculture incubator to provide land and appropriate infrastructure for urban agriculture enterprises
Philadelphia	Green Plan Philadelphia (forthcoming) www.greenplanphiladelphia.com	The final version of this plan will include a chapter on urban agriculture and food access.
Alexandria, Va.	Eco-City Alexandria Environmental http://ecocity.ncr.vt.edu/ actionplan.html	Encourages local and regional food production and equitable access to safe, healthy, and organic food, for children and adolescents in particular

Appendix 4. Urban Agriculture Components in Regional Plans

Jurisdiction	Plan Name, Date, and Link	community gardens	institutional gardens	private gardens	edible landscaping	commercial gardens	commercial farms	rooftop urban agriculture	orchards	bees	chickens	farm animals	growing space	uncontaminated soil	land tenure	water	financial assistance	agricultural practices	agricultural skills and knowledge	environmental stewardship	community development	health/nutrition education	other topics	rural agriculture	food processing	food distribution	food retail	food access and availability	food literacy and education	food waste and disposal	community engagement
																				Urban Agriculture Topics Addressed by the Plan								*Other Food System Topics*			
Capitol Region Council of Governments (Hartford, Conn.)	Regional Plan of Conservation and Development: Chapter 6, Food System, 2009 www.crcog.org/community_dev/regional_plan.html						×											×						×	×	×		×		×	
Waterloo Region (Ontario)	A Healthy Community Food System Plan for Waterloo Region, 2007 www.region.waterloo.on.ca/web/health.nsf/vwSiteMap/54ED787F44ACA44C852571410056AEB0/$file/FoodSystem_Plan.pdf?openelement				×								×								×				×	×	×	×		×	×
Waterloo Region (Ontario)	The Regional Official Plan (draft), April 2009 www.region.waterloo.on.ca/web/region.nsf/DocID/CA5BC18540AE6A2185257555006D0304	×			×																							×			
Delaware Valley Regional Planning Commission (Philadelphia)	Connections: The Regional Plan for a Sustainable Future, 2009 www.dvrpc.org/Connections	×		×				×					×			×				×	×		×		×			×			×

(continued)

118

Appendix 4. Urban Agriculture Components in Regional Plans

Jurisdiction	Plan Name, Date, and Link	How does plan support urban agriculture?
Capitol Region Council of Governments (Hartford, Conn.)	Regional Plan of Conservation and Development: Chapter 6, Food System, 2009 www.crcog.org/community_dev/regional_plan.html	Strong support for rural agriculture and local retailing
Waterloo Region (Ontario)	A Healthy Community Food System Plan for Waterloo Region, 2007 www.region.waterloo.on.ca/web/health.nsf/vwSiteMap/54ED787F44ACA44C8525714I0056AEB0/$file/FoodSystem_Plan.pdf?openelement	Encourages community engagement and training, gardens, and food access as ways to promote urban agriculture
Waterloo Region (Ontario)	The Regional Official Plan (draft), April 2009 www.region.waterloo.on.ca/web/region.nsf/DocID/CA5BC18540AE6A2185257555006D0304)	Promotes community gardens and tree planting as major urban greenlands initiatives
Delaware Valley Regional Planning Commission (Philadelphia)	Connections: The Regional Plan for a Sustainable Future, 2009 www.dvrpc.org/Connections	Identifies urban agriculture as critical element of a community's green infrastructure. Encourages gardens, community engagement, and water quality protection as measures to implement urban agriculture

Appendix 5. Urban Agriculture–Related Zoning Regulations

Urban agriculture activities acknowledged as permitted uses in multiple existing zoning districts

Jurisdiction	Code citation	How does this support urban ag?
Elk Grove, Calif.	Municipal Code. Title 23: Zoning Code. Chapter 23.30: Residential Zoning Districts. Section 23.30.32: Allowed Uses and Permit Requirements (www.codepublishing.com/CA/elkgrove)	Permits community gardens and livestock in all residential districts. Permits 2 livestock animals for every 1/2 acre of land
Glendale, Calif.	Code of Ordinances. Title 30: Zoning. Chapter 30.11: Residential Districts; Chapter 30.12: Commercial Districts; Chapter 30.14: Mixed Use Districts (http://library.municode.com/index.aspx?clientId=16369&stateId=5&stateName=California)	Permits community gardens run by home owners associations by right in all residential districts and community gardens run by nonprofits by right in all commercial and mixed use districts
Los Angeles	Municipal Code. Chapter I: Planning and Zoning Code. Article 2: Specific Planning - Zoning Comprehensive Zoning Plan (www.amlegal.com/nxt/gateway.dll?f=templates&fn=default.htm&vid=amlegal:lapz_ca)	Permits truck gardening and poultry in a number of residential districts
San Francisco	Planning Code. Article 2: Use Districts. Section 227: Other Uses (http://library.municode.com/index.aspx?clientId=14139&stateId=5&stateName=California)	Permits truck gardens in a number of commercial and manufacturing districts
Santa Cruz, Calif.	Municipal Code. Title 24: Zoning. Chapter 24.10: Land Use Districts (www.codepublishing.com/CA/SantaCruz)	Permits community gardens in a number of residential districts and permits commercial agriculture with an administrative use permit in two industrial districts
Denver	Zoning Code. Article 11: Use Limitations. Section 11.6.1: Garden, Urban; Section 11.8.4: Garden; Section 11.8.6: Keeping of Household Animals; Section 11.10.9: Garden (www.denvergov.com/cpd/ZoningCodeMapWhatsMyZoning/tabid/432507/Default.aspx)	Permits urban gardens (with accessory structures and beekeeping) as primary uses in most districts, subject to administrative review and specific standards. Permits on-site sales for urban gardens in nonresidential districts. Permits accessory gardens in all residential districts, subject to specific standards. Permits accessory beekeeping in most districts, subject to specific standards. Permits accessory gardens in most nonresidential districts, subject to specific standards
Fort Collins, Colo.	Land Use Code. Article 3: General Development Standards. Division 3.8: Supplementary Regulations. Section 3.8.1: Accessory Buildings, Structure, and Uses. Also see Article 4: Districts (www.colocode.com/ftcollins/landuse/begin.htm)	Permits "cultivation, storage and sale of crops, vegetables, plants and flowers produced on the premises" as an accessory use in most zoning districts, "when the facts, circumstances and context of such uses reasonably so indicate"
Washington, D.C.	Municipal Regulations. Title 11: Zoning. Chapter 11-2: R-1 Residence District Use Regulations; Chapter 11-3: R-2, R-3, R-4, and R-5 Residence District Use Regulations; Chapter 11-5: Special Purpose Districts (www.dcregs.org/Gateway/TitleHome.aspx?TitleNumber=11)	Permits farms and truck gardens by right in all residential districts

(continued)

Jurisdiction	Citation	Provision
Safety Harbor, Fla.	Comprehensive Zoning & Land Development Code. Article III: Supplementary District Regulations. Section 41.00: Community Gardens (www.cityofsafetyharbor.com/DocumentView.aspx?DID=1436)	Permits community gardens in most districts, subject to specific development standards
St. Petersburg, Fla.	City Code. Chapter 16: Land Development Regulations. Section 16.50.085: Community Gardens (http://library.municode.com/index.aspx?clientId=11602&stateId=9&stateName=Florida)	Permits community gardens in all zoning districts, subject to specific standards
Nampa, Idaho	City Code. Title 10: Planning and Zoning. Chapter 3: Establishment of Districts and Provisions for Nonconforming Uses. Section 10-3-2: Schedule of District/Zone Land Use Controls. Also see Chapter 21: Animal Zoning Regulations (http://ci.nampa.id.us/pages/department-portal.php?deptid=21)	Permits gardening for personal consumption in all residential districts and commercial agriculture in a number of nonresidential districts. Permits apairies on lots of any size and livestock on lots of at least 30,000 square feet
New Orleans	Comprehensive Zoning Ordinance. Article 4: Residential Districts; Article 5: Business and Commercial Districts; Article 6: Central Business Districts (http://library.municode.com/index.aspx?clientId=16306&stateId=18&stateName=Louisiana)	Permits farms on sites of at least 5 acres and private and truck gardens (without on-site sales) by right in all residential and commercial districts
Bloomington, Ind.	Code of Ordinances. Title 20: Unified Development Ordinance. Chapter 20.02: Zoning Districts. Also see Chapter 20.05: Development Standards. Section 20.05.092: SC-07 Special Conditions; Crops and Pasturage, and Accessory Chicken Flocks (http://bloomington.in.gov/code)	Permits community gardens by right in residential districts. Permits the keeping of poultry and livestock in low-density residential districts, subject to specific standards
Fall River, Mass.	Code of Ordinances. Chapter 86: Zoning. Article III: Districts and District Use Regulations (http://library5.municode.com/default-now/home.htm?infobase=14774&doc_action=whatsnew)	Permits farms and market gardens by right in a number of higher-density residential districts and in a number of business districts outside of the CBD
Duluth, Minn.	Unified Development Code. Article 3: Permitted Uses. Table 50-19-8: Use Table (http://www.duluthmn.gov/planning/zoning_regulations/index.cfm)	Permits urban agriculture by right in most residential districts
Grand Rapids, Minn.	Code of Ordinances. Title V: Zoning and Planning. Chapter 61: Zoning Ordinance. Article 5: Residential Zone Districts. Section 5.5.05: Uses of Land. Also see Article 6: Mixed-Use Commercial Zone Districts. Section 5.6.06: Uses of Land (http://library.municode.com/index.aspx?clientId=12116&stateId=22&stateName=Michigan)	Permits community gardens by right in most zoning districts
Minneapolis	Code of Ordinances. Title 20: Zoning Code. Chapter 536: Specific Development Standards. Section 536.20: Specific Development Standards: Community Garden (www.municode.com/resources/gateway.asp?pid=11490&sid=23)	Permits community gardens in all but two zoning districts, subject to specific development standards
Kansas City, Mo.	Second Committee Substitute for Ordinance No. 100299, As Amended (www.kcmo.org/idc/groups/cityplanningdevelopmentdiv/documents/cityplanninganddevelopment/1 00299.pdf)	Permits home gardens, community gardens, and CSA farms in almost all zoning districts subject to specific use standards. Permits on-site sales for community gardens and CSAs
Springfield, Mo.	Zoning Ordinance. Section 5-3000: Community Gardens (www.springfieldmo.gov/zoning/pdfs/ZO_032210.pdf)	Permits community gardens in single-family residential districts subject to specific development standards

(continued)

City	Code Reference	Provisions
Cary, N.C.	Code of Ordinances. Appendix A: Land Development Ordinance. Chapter 5: Use Regulations. Section 5.1: Tables of Permitted Uses (www.amlegal.com/library/nc/cary.shtml)	Permits community gardens by right in most zoning districts
Charlotte, N.C.	Code of Ordinances. Appendix A: Zoning. Chapter 9: General Districts (http://library.municode.com/index.aspx?clientId=19970&stateId=33&stateName=North Carolina)	Permits farms with on-site sales by right in all residential, business, and institutional districts as long as the farm site is a minimum of 3 acres
Greensboro, N.C.	Code of Ordinances. Chapter 30: Zoning, Planning, and Development Ordinance. Article IV: Zoning. Section 30-4-5.1: Permitted Uses (www.municode.com/Library/clientCodePage.aspx?clientID=12033)	Permits crop production by right in all but one zoning district
Raleigh, N.C.	Code of Ordinances. Part 10: Planning and Development. Article D: Section 10-2071: Schedule of Permitted Land Uses in Zoning Districts (http://library.municode.com/index.aspx?clientId=10312&stateId=33&stateName=North Carolina)	Permits general agriculture (raising crops and animals) by right in most nonresidential districts. Permits research farms in most districts by right
New York City	Zoning Resolution. Article II: Residence District Regulations. Chapter 2: Use Regulations. Section 22-10: Uses Permitted As of Right. Also see Article IV: Manufacturing District Regulations. Chapter 2: Use Regulations. Section 42-10: Uses Permitted As of Right (www.nyc.gov/html/dcp/html/zone/zonetext.shtml)	Permits agricultural uses including greenhouses and truck gardens by right in all residential districts and in multiple manufacturing districts
Cincinnati	Code of Ordinances. Title XIV: Zoning Code. Section 1419-41: Community Gardens (http://library.municode.com/index.aspx?clientId=19996&stateId=35&stateName=Ohio)	Permits community gardens by right in most zoning districts, subject to specific use standards
Cleveland	Codified Ordinances. Part 3 Land Use Code - Planning and Housing. Title VII: Zoning Code. Chapter 347: Specific Uses Regulated. Section 347.02: Restrictions on the Keeping of Farm Animals and Bees (http://caselaw.lp.findlaw.com/clevelandcodes/cco_part3_347.html)	Permits the keeping of poultry, livestock, and bees in residential and nonresidential districts subject to specific standards
Toledo, Ohio	Municipal Code. Part 11: Planning and Zoning Code. Chapter 1104: Use Regulations. Section 1104.0100: Use Table. Also see Chapter 1116: Terminology. Section 1116.0202: Agriculture (www.ci.toledo.oh.us/Portals/0/Planning%20Docs/Toledo%20Planning%20and%20Zoning%20Code.pdf)	Permits agricultural activities either by right or with a special permit in most zoning districts
Lincoln City, Ore.	Municipal Code. Title 17: Zoning. Chapter 17.80: Provisions Applying to Special Uses. Section 17.80.080: Animals and Gardens (www.codepublishing.com/OR/LincolnCity)	Permits 5 domestic fowl on any lot. Permits farm animals on any lot of at least 20,000 square feet, with a minimum of 10,000 square feet per animal. Permits personal gardens with related hoop houses or greenhouses as accessory uses on any lot. Permits community gardens and market gardens with related hoop houses, greenhouses, or tool sheds in a number of residential and commercial districts and special planning areas
Portland, Ore.	City Code. Title 33: Zoning Code. Chapter 33.100: Open Space Zone; Chapter 33.110: Single Family Dwelling Zones; Chapter 33.130: Commercial Zones; Chapter 33.140: Employment and Industrial Zones. Also see Chapter 33.920: Description of Use Categories. Section 33.920.500: Agriculture (www.portlandonline.com/auditor/index.cfm?c=28197)	Permits agricultural activities by right in certain zoning districts and subject to conditional use review in others

(continued)

Jurisdiction	Code Citation	Summary
Providence, R.I.	Code of Ordinances. Chapter 27: Zoning. Article III: Use and Dimensional Regulations. Section 303: Use Regulations. Also see Article XI: Amendments and Validity. Appendix A (www.municode.com/Library/clientCodePage.aspx?clientID=12107)	Permits community gardens, crop farming, and truck gardens in most zoning districts by right
Memphis/Shelby County, Tenn.	Unified Development Code. Article 2: Districts and Uses. Section 2.5: Permitted Use Table; Section 2.6.1.4: Live chickens (hens only) kept on a single family detached lot); Section 2.6.3.Q: Neighborhood Gardens (http://memphis.code-studio.com/PDF/UDC_6-10-10.pdf)	Permits neighborhood gardens in most zoning districts by right, subject to standards. Permits farmers markets in multiple nonresidential districts by right, subject to standards. Permits up to 6 hens on single-family residential lots with a health permit subject to standards
Nashville-Davidson, Tenn.	Code of Ordinances. Title 17: Zoning. Chapter 17.08, Section 17.08.030: District Land Use Tables. Chapter 17.16, Section 17.16.230.A: Commercial Community Garden (http://library6.municode.com/default-test/home.htm?infobase=14214&doc_action=whatsnew)	Permits noncommercial community gardens by right in most zoning districts and permits commercial community gardens as special exception uses in numerous residential districts, subject to standards
Austin, Tex.	City Code. Title 25: Land Development. Chapter 25-2: Zoning. Subchapter C, Article 4, Division 4, Section 25-2-863: Urban Farms (www.amlegal.com/library/tx/austin.shtml)	Permits farms of 1 to 5 acres in a wide range of districts, subject to standards
Dallas	City Code. Volume III, Chapter 51A, Part II, Article IV: Zoning Regulations. Division 51A-4.200: Use Regulations. Section 51A-4.201: Agricultural Uses (/www.amlegal.com/library/tx/dallas.shtml)	Permits crop production in most zoning districts, subject to specific development standards
Waco, Tex.	Code of Ordinances. Chapter 28: Zoning. Article V: Supplementary District Regulations. Division 7: Community Gardens (http://library7.municode.com/default-test/home.htm?infobase=11666&doc_action=whatsnew)	Permits community gardens (with limited on-site sales) in any zoning district as special uses, subject to specific standards and a special use review process. The city views community gardens as a vacant-land management tool.
Roanoke, Va.	Code of Ordinances. Chapter 36.2: Zoning. Article 3: Regulations for Specific Zoning Districts. Section 36-2-340: Use Matrix (http://library.municode.com/index.aspx?clientId=11474&stateId=46&stateName=Virginia)	Permits community gardens by right in all zoning districts
Burlington, Vt.	Code of Ordinances. Appendix A: Comprehensive Development Ordinance. Appendix A: Use Table--All Zoning Districts (http://library4.municode.com/default-test/home.htm?infobase=13987&doc_action=whatsnew)	Permits community gardens in almost all districts by right
Bothell, Wash.	Municipal Code. Title 12: Zoning. Chapter 12.06: Permitted Uses. Section 12.06.030: Agriculture (www.codepublishing.com/wa/bothell)	Permits crop production with on-site sales in all zoning districts and animal keeping in all residential zoning districts, subject to development standards
Des Moines, Wash.	Municipal Code. Title 18: Zoning. Chapter 18.33: Keeping of Animals in Residential Zones (www.codepublishing.com/wa/desmoines)	Permits 10 fowl per 22,000 square feet on single-family lots, plus an additional 5 fowl for each 11,000 square feet of lot size. Permits one larger farm animal on residential lots of at least 35,000 square feet, with an allowance for an additional animal for every additional 17,500 square feet of lot size

(continued)

123

Jurisdiction	Citation	Description
Issaquah, Wash.	Municipal Code. Title 18: Land Use Code. Chapter 18.07: Required Design and Development Standards. Section 18.07.140: Animals - Maintenance of Agricultural Animals in Residential District; Section 18.07.160: Honey Bees (www.codepublishing.com/wa/issaquah)	Permits 1 hen per 2,000 square feet of lot area for lots between 6,000 and 35,000 square feet and permits 1 livestock animal per 1/2 acre on lots over 35,000 square feet. Hen enclosures must be 5 feet from property lines. Larger animals must be kept at least 25 feet from property lines.
Port Townsend, Wash.	Municipal Code. Title 17: Zoning. Chapter 17.16: Residential Zoning Districts. Table 17.16.120: Residential Zoning Districts – Permitted, Conditional and Prohibited Uses. Also see Chapter 17.24: Public, Park, and Open Space Zoning Districts. Table 17.24.020: Public, Park and Open Space Zoning Districts – Permitted, Conditional and Prohibited Uses (www.codepublishing.com/WA/PortTownsend.html)	Permits crop production, CSA farms, and noncommercial animal husbandry in low-density residential districts. Permits "community agriculture centers" (commercial farms) as conditional uses in low-density residential districts. Permits community gardens in all open space districts and CSA farms in Public/Infrastructure districts. Permits community agriculture centers as conditional uses in Public/Infrastructure districts
Seattle	Council Bill 116907 (http://clerk.ci.seattle.wa.us/~scripts/nph-brs.exe?s1=&s3=116907&s4=&s2=&s5=&Sect4=AND&l=20&Sect2=THESON&Sect3=PLURON&Sect5=CBORY&Sect6=HITOFF&d=ORDF&p=1&u=%2F%7Epublic%2Fcbory.htm&r=1&f=G)	Permits community gardens and urban farms in all zoning districts, subject to specific conditions. Also addresses chickens, rooftop greenhouses, and farmers markets
Seattle	Municipal Code. Title 23: Land Use Code. Subtitle III: Land Use Regulations. Division 2: Authorized Uses and Development Standards. Chapter 23.42: General Use Provisions. Section 23.42.052: Keeping of Animals (http://clerk.ci.seattle.wa.us/~scripts/nph-brs.exe?d=CODE&s1=23.42.052.snum.&Sect5=CODE1&Sect6=HITOFF&l=20&p=18&u=/~public/code1.htm&r=1&f=G)	Permits the keeping of poultry, livestock, and bees in all districts, subject to specific standards
Brown Deer, Wis.	Code of Ordinances. Subpart B: Land Development Regulations. Chapter 121: Zoning. Article IV: Residence Districts; Article V: Business Districts (http://library6.municode.com/default-test/home.htm?infobase=13877&doc_action=whatsnew)	Permits community gardens by right in all residential and most business districts
Madison, Wis.	Code of Ordinances. Chapter 28: Zoning Code. Section 28.08: Residence Districts (http://library.municode.com/index.aspx?clientId=50000&stateId=49&stateName=Wisconsin)	Permits 4 hens in low-density residential districts, provided they are kept in a covered pen at least 25 feet from neighboring residences. Requires a license
Milwaukee	Zoning Ordinance. Subchapter 5: Residential Districts. Table 295-503-1: Residential Use Districts Table (www.mkedcd.org/czo)	Permits raising of crops or livestock by right in all residential districts

Urban agriculture activities acknowledged as permitted uses in multiple existing zoning districts (proposed amendments as of 7/31/2010)

Jurisdiction	Code citation	How does this support urban ag?
Baltimore	Draft Zoning Code. Title 14: Use Standards. Section 14-305: Community Garden (permanent); Section 14-327: Urban Agriculture (proposed; (www.transformbaltimore.net/portal/zoning-apr-draft?tab=files)	Baltimore's proposed zoning permits community gardens (no animal husbandry, heavy equipment, or processing) with farm stands, subject to standards, in all residential districts and many nonresidential districts. The proposed code also permits urban agriculture (including animal husbandry, heavy equipment, and processing) with farms stands conditionally in most zoning districts, subject to standards.
Philadelphia	Zoning Code. Module 2: Zoning Districts and Uses (proposed; www.zoningmatters.org/files/Public_Draft_Module_2_-_Districts_and_Uses.pdf)	Philadelphia's draft zoning code permits community gardens, market farms, and CSA farms in a wide range of residential and nonresidential districts. Animal husbandry would be permitted in two industrial districts.

Special urban agriculture districts

Jurisdiction	Code citation	How does this support urban ag?
Owensboro, Ky.	Zoning Ordinance. Article 8: Development Zones. Section 8.11: Agriculture Zones; Section 8.2: Zones and Use Table; Section 8.4: Detailed Uses and Special Conditions of Zones and Use Table; Section 8.5: Site Development (www.iompc.com/documents/Z_PDFS/z08.pdf)	Permits the raising of crops and livestock and on-site sales of crops on lots of at least 1/2 acre
Boston	Zoning Code. Chapter 33: Open Space Subdistricts (www.cityofboston.gov/bra/pdf/ZoningCode/Article33.pdf)	Permits community gardens to be designated as open space subdistricts
Livonia, Mich.	Zoning Ordinance. Article V: R-U-F District Regulations (http://library1.municode.com/default-now/home.htm?infobase=13644&doc_action=whatsnew)	The "Rural Urban Farm" district permits the growing and selling of vegetables on lots of at least 1/2 acre
Cleveland	Codified Ordinances. Part Three, Title VII: Zoning Code. Chapter 336: Urban Garden District (http://caselaw.lp.findlaw.com/clevelandcodes/cco_part3_336.html)	Permits community gardens and market gardens and includes specific allowances for accessory structures and on-site sales
Chattanooga, Tenn.	City Code. Appendix B: Zoning Regulations. Article V: Zone Regulations. Section 1600: A-1 Urban Agricultural Zone (www.chattanooga.gov/city_council/code/41%20--%20code%20appendix%20b%20-%20zoning%20ordinance.pdf)	Permits a wide range of agricultural activities including raising crops and livestock and on-site sales of crops on lots of at least 20 acres
Burlington, Vt.	Code of Ordinances. Appendix A: Comprehensive Development Ordinance. Article 4: Zoning Maps and Districts. Part 4: Base Zoning District Regulations. Section 4.4.3: Enterprise Districts (http://library4.municode.com/default-test/home.htm?infobase=13987&doc_action=whatsnew)	The Agricultural Processing and Energy district, a hybrid agricultural/light industrial district just north of downtown, permits a wide range of agricultural activities including commercial crop production and livestock. The purpose statement for the district encourages businesses to build linkages to improve efficiency.

Special urban agriculture districts (proposed as of July 31, 2010)

Jurisdiction	Code citation	How does this support urban ag?
Madison, Wis.	Zoning Ordinance. Chapter 28G: Special Districts (proposed; www.cityofmadison.com/neighborhoods/zoningrewrite/documents/Special_Districts.pdf)	Madison's proposed zoning adds a Special Urban Agriculture District to address small-scale agricultural uses within city limits.

(continued)

Urban agriculture activities acknowledged in form-based codes or in PUD, TND, or conservation subdivision development options

Jurisdiction	Code citation	How does this support urban ag?
Jacksonville, Ala.	Code of Ordinances. Chapter 24: Zoning. Division 5: Special Districts. Section 24-173: PUD Planned Unit Development (http://library.municode.com/index.aspx?clientId=14051&stateId=1&stateName=Alabama)	Encourages the provision of community garden space as an energy efficiency measure in PUDs
Chico, Calif.	Municipal Code. Title 19: Land Use and Development Regulations. Division VI: TND Regulations (www.chico.ca.us/document_library/municode/Title19.pdf)	Permits commercial crop production as well as animal keeping, subject to standards, in the Neighborhood Edge subdistrict. Lists community gardens as a permitted open-space use in multiple subdistricts
Milford, Del.	Code of Ordinances. Part II: General Legislation. Chapter 230: Zoning. Article III: Use and Area Regulations. Section 230-9.C.11: Planned Residential Neighborhood Development (http://library.municode.com/index.aspx?clientId=14818&stateId=8&stateName=Delaware)	Permits a 5% density bonus for Planned Residential Neighborhood Development projects that set aside additional open space for community gardens
Miami	Miami 21 Code (www.miami21.org/final_code_AsAdoptedMay2010.asp)	Permits community gardens as part of required open space in most transects. Permits community gardens in Civic Use zones and as a public benefit for developers seeking increased height or floor area
Atlanta	Code of Ordinances. Part III: Land Development Code. Part 16: Zoning. Chapter 19E: PD-CS Planned Development—Conservation Subdivision District Regulations	Permits community gardens and composting areas within required green spaces for conservation subdivisions
Forsyth, Ga.	Zoning Ordinance. Article 9: TND, Traditional Neighborhood Development District (www.cityofforsyth.net/docs/Zoning%20Ordiance/9%20Traditional%20Neighborhood%20Development%20District.pdf).	Permits noncommercial gardens as accessory uses for single-family homes in TND districts
Urbana, Ill..	Zoning Ordinance. Article XIII: Special Development Provisions. Section XIII-3: Planned Unit Developments (http://urbanaillinois.us/sites/default/files/attachments/2008_Zoning_Ordinance.pdf).	Lists community gardens as a recommended passive recreation feature for PUDs
Denmark Township, Minn.	Development Code. Chapter 2: Zoning Regulations. Part 3: Performance Standards. Section 4: Open Space Design (www.denmarktownship.org/vertical/Sites/%7B5B1C76D5-2975-4BD2-B0BA-7327C34C7928%7D/uploads/%7BCA5565C1-667B-43F1-8718-1F718B6B1027%7D.PDF).	Permits demonstration farms, community gardens, and composting as permitted open-space uses within conservation subdivisions
Mint Hill, N.C.	Code of Ordinances. Appendix A: Zoning. Section 7.2: Downtown Mint Hill Overlay Code (www.minthill.com/documents/Downtown/Downtown%20Overlay%20Code%2001-05-10.PDF)	Form-based code that lists community gardens as a permissible open-space type in the Neighborhood transect
North Kingston, R.I.	Revised Ordinances. Chapter 21: Zoning. Article IX: Conservation Developments (http://library4.municode.com/default-test/home.htm?infobase=11995&doc_action=whatsnew)	Permits crop production with accessory on-site sales on open space lands in conservation subdivisions
North Myrtle Beach, S.C.	Code of Ordinances. Chapter 23: Zoning. Article II: Zoning Districts and Development Regulations. Section 23.29: PDD Planned Development District (http://library6.municode.com/default-test/home.htm?infobase=11359&doc_action=whatsnew)	Lists community gardens and organic composting sites as uses that demonstrate environmental stewardship in the site design of Planned Development District projects
Hutto, Tex.	Hutto SmartCode (www.gatewayplanning.com/PDFS/Hutto,Texas%20SmartCode.pdf)	Form-based code that permits a wide variety of different food production uses in different transects

(continued)

| Chesapeake, Va. | Code of Ordinances. Appendix A: Zoning. Article 6: Residential Districts. Section 6-2200: Residential Cluster Development Standards (http://library1.municode.com/default-test/home.htm?infobase=10529&doc_action=whatsnew) | Permits agricultural operations (crop production and animal husbandry) and community gardens on conservation lands in conservation subdivisions |

Farmers markets permitted in multiple zoning districts

Jurisdiction	Code citation	How does this support urban ag?
Durham, N.C.	Unified Development Ordinance. Article 5. Section 5.1.2: Use Table; Section 5.2.5.F: Retail Sales and Service (www.durhamnc.gov/udo)	Outdoor farmers markets are listed as a primary use in the retail sales and service use category. Farmers markets are permitted by right or with a development plan in a number of mixed use and commercial districts.
Gainesville, Fla.	Code of Ordinances. Chapter 30: Land Development Code. Article VI: Requirements for Specially Regulated Uses. Section 30-115: Farmers Markets (http://library.municode.com/index.aspx?clientId=10819&stateId=9&stateName=Florida)	Farmers markets are permitted on public or private property with a special permit and subject to specific conditions.
Grand Junction, Colo.	Zoning and Development Code. Chapter 21.04: Uses. Section 21.04.010: Use Table (www.ci.grandjct.co.us/CityDeptWebPages/CommunityDevelopment/DevelopmentServices/PDF/ZoningAndDevelopmentCode/ZoningCodeFinalCLEAN2010April05.pdf)	Farmers markets are permitted uses in a number of commercial and mixed use districts.
Grand Rapids, Mich.	Code of Ordinances. Title V: Zoning and Planning. Chapter 61: Zoning Ordinance. Article 9: Use Regulations. Section 5.9.32.J.2: Farmers Markets (http://library.municode.com/index.aspx?clientId=12116&stateId=22&stateName=Michigan)	Farmers markets are permitted as temporary uses (maximum of 9 months per year) in mixed use commercial districts, subject to standards.
Little Elm, Tex.	Code of Ordinances. Chapter 106: Zoning. Article I: In General. Section 106-33.5: Farmers Market Regulations (http://library.municode.com/index.aspx?clientId=13870&stateId=43&stateName=Texas)	Farmers markets are permitted uses in multiple zoning districts, subject to specific operational and site standards.
Madison, Wis.	Code of Ordinances. Chapter 28: Zoning Code. Section 28.08: Residence Districts; Section 28.085: Office Districts; Section 28.09: Commercial Districts; Section 28.10: Manufacturing Districts. Also see Section 28.12: Administration and Enforcement. Part 28.12(11): Conditional Uses (http://library.municode.com/index.aspx?clientId=50000&stateId=49&stateName=Wisconsin)	Farmers markets are permitted as conditional uses in the parking lots of nonresidential uses in R-1, R-4L, R-4A, and R-1R districts. Farmers markets are permitted by right in the parking lots of nonresidential uses in O-3, O-4, C-2, and M-1 districts.
Minneapolis	Code of Ordinances. Title 20: Zoning Code. Chapter 535. Article V: Temporary Uses. Section 535.360: Permitted Temporary Uses and Structures. Also see Chapter 536. Section 536.20: Specific Development Standards (http://library1.municode.com/default-test/home.htm?infobase=11490&doc_action=whatsnew)	Farmers markets are permitted as temporary uses in all but one industrial zoning district. Permanent farmers markets are permitted, subject to standards, in most nonresidential districts.

(Note: standout examples are highlighted in yellow.)

Appendix 6. Allowances for Poultry, Livestock, or Bees in Animal Control Ordinances

Jurisdiction	Code citation	How does this support urban ag?
Mobile, Ala.	Code of Ordinances. Chapter 7: Animals and Fowl. Article IV: Livestock and Poultry (http://library.municode.com/index.aspx?clientId=11265&stateId=1&stateName=Alabama)	Allows 25 hens with permit as long as enclosures are kept at least 40 feet from neighboring residences and 20 feet from property lines. Allows a cow with permit with a corral of at least 12,000 square feet as long as the corral is at least 150 feet from neighboring residences
Rogers, Ark.	Ordinance No. 06-100 (www.rogersarkansas.com/clerk/chkordinance.asp)	Allows 4 hens with permit in single-family districts as long as the enclosure is kept at least 25 feet from neighboring residences
Mountain View, Calif.	Code of Ordinances. Chapter 5: Animals and Fowl. Article II: Miscelleanous Regulations (http://library.municode.com/index.aspx?clientId=16508&stateId=5&stateName=California)	Allows 4 hens without permit as long as their enclosure is at least 25 feet from neighboring residences. Allows 2 livestock on lots of 1 acre with permit
Morgan Hill, Calif.	Municipal Code. Title 6: Animals. Chapter 6.36: Animals and Land Use. Section 6.36.175: Keeping of Livestock for Private Uses in Residential Zoning Districts and Open Space Properties with a Private Residence (www.municode.com/Resources/gateway.asp?pid=16502&sid=5)	Allows both large and small livestock animals with permit in residential districts subject to specific conditions
Santa Clara, Calif.	City Code. Title 6: Animals. Chapter 6.15: Keeping of Certain Animals (www.codepublishing.com/ca/santaclara)	Allows an unspecified number of hens, roosters, and other large farm animals as long as they comply with the distancing and performance standards in Chapter 6.15. Hens must be kept at least 50 feet from neighboring dwelling units. Roosters must be kept 100 feet from neighboring dwelling units, and large farm animals must be kept at least 100 feet from all dwelling units. Exceptions require a permit.
Fort Collins, Colo.	Municipal Code. Chapter 4: Animals and Insects. Article II: Animals. Division 6: Restrictions. Also see Article III: Insects. Division 2: Bees (www.colocode.com/ftcollins/municipal/begin2.htm#toc)	Allows 6 hens per parcel of property with permit from Humane Society as long as enclosures are kept at least 15 feet from property lines. Allows 2 bee hives on lots 1/4 acre and smaller. Allows additional hives on larger lots
Longmont, Colo.	Ordinance O-2009-05. A Bill for an Ordinance Amending Chapter 7.04 Animals of the Longmont Municipal Code Regarding Backyard Chicken Hens (www.ci.longmont.co.us/city_clerk/muni_code/documents/O-2009-05.pdf)	Allows up to 4 hens with permit in residential backyards as long as chicken coops are kept at least 7 feet away from rear and side lot lines. Permits capped at 50
South Portland, Me.	Code of Ordinances. Chapter 3: Animals and Fowl. Article II: Domesticated Chickens. Article III: Beekeeping (www.southportland.org/vertical/Sites/%7B7A5A2430-7EB6-4AF7-AAA3-59DBDCFA30F2%7D/uploads/%7BDEFCD6F9-B7DA-4D71-9DC5-B897EC37DA0D%7D.PDF)	Allows 6 hens per residential lot with permit and noncommercial beekeeping subject to specific conditions. Chicken permits capped at 20 per year
Ann Arbor, Mich.	City Code. Title IX. Chapter 107: Animals. Section 9.39: Bees; Section 9.42: Keeping of Chickens (http://library1.municode.com/default-now.htm?infobase=11782&doc_action=whatsnew)	Allows 4 hens with permit in low-density residential districts as long as enclosures are located at least 10 feet from property lines and 40 feet from neighboring dwellings. Allows 2 beehives
Kansas City, Mo.	Code of Ordinances. Chapter 14: Animals. Section 14-12: Keeping of Livestock Generally; Keeping of Wild Animals; Section 14-15: Keeping of Small Animals and Fowl in Pens (http://library3.municode.com/default-test/home.htm?infobase=10156&doc_action=whatsnew)	Allows 2 large animals as long as they are kept at least 200 feet away from neighborhing homes. Allows 15 chickens as long as they are kept at least 100 feet away from neighboring homes and 1 rooster if it is kept 300 feet from neighboring homes. Requires permits for animals

(continued)

City	Ordinance	Description
Missoula, Mont.	Municipal Code. Title 6: Animals. Chapter 6.12: Keeping Livestock and Fowl (www.ci.missoula.mt.us/DocumentView.aspx?DID=1028#Chapter_6_12)	Allows 6 hens with permit on single-family residential lots provided the chickens are kept at least 20 feet away from adjacent residential structures
Round Rock, Tex.	Code of Ordinances. Chapter 2: Animal Control. Section 2.700: Livestock; Section 2.800: Fowl (www.roundrocktexas.gov/docs/ordinances_ch_02.pdf)	Allows 1 farm animal on lots of at least 1 acre with 1 additional animal for each 1/2 acre of lot size. Allows 10 chickens as long as enclosures are kept 50 feet from neighboring building and allows 5 chickens if enclosure is between 25 and 50 feet from neighboring buildings
San Antonio, Tex.	Code of Ordinances. Chapter 5: Animals. Article III: Livestock. Section 5-52: Keeping of Bovines, Equines, Sheep, and Goats. Also see Article V: Animal Licenses and Permits. Section 5-109: Animal Limits, Excess Animal Permit; Section 5-114: Livestock Permits (www.municode.com/resources/gateway.asp?pid=11508&sid=43)	Allows 3 domestic fowl and 2 livestock animals with permit subject to specific conditions

(Note: This list only highlights a few of hundreds of animal control ordinances from across the country.)

Appendix 7. Other Municipal Policies Supporting Urban Agriculture

JURISDICTION	TYPE OF REGULATION OR POLICY	CODE CITATION	HOW DOES THIS SUPPORT URBAN AGRICULTURE?
Composting regulations			
Champaign, Ill.	Public health ordinance	Municipal Code. Chapter 35: Vegetation. Article V: Composting (www.municode.com/Resources/gateway.asp?pid=10520&sid=13)	Includes specific location, performance, a materials standards for compost piles
Chicago	Public health ordinance	Municipal Code. Title 7: Health and Safety. Chapter 7-28: Health Nuisances. Article V: Rat Control. Section 7-28-710: Dumping Prohibited; Section 7-28-715: Composting Standards (www.amlegal.com/nxt/gateway.dll/Illinois/chicago_il/municipalcodeofchicago?f=templates$fn=default.htm$3.0$vid=amlegal:chicago_il)	Section 7-28-710 clarifies the city's pro-composting stance. Section 7-28-715 dete standards to ensure that compost piles dc become public nuisances.
North Kansas City, Mo.	Public health ordinance	Municipal Code. Title 8: Health and Safety. Chapter 8.32: Solid Waste Collection and Disposal. Section 8.32.100: Composting of Yard Wastes (www.municode.com/Resources/gateway.asp?pid=16014&sid=25)	Includes specific location, performance, a materials standards for compost piles
Burnsville, Minn.	Public health ordinance	City Code. Title 7: Health and Sanitation. Chapter 10: Composting (www.sterlingcodifiers.com/codebook/index.php?book_id=468)	Includes specific location, performance, a materials standards for compost piles
Platting and impact fee exemption for urban agriculture			
Austin, Tex.	Subdivision ordinance	Austin City Code. Title 8: Parks and Recreation. Chapter 8-4: Qualified Community Garden. See also Title 25: Land Development. Chapter 25-4. Subdivision. Article 1: Subdivision Compliance. Section 25-4-3: Temporary Exemption from Platting Requirements; also see Chapter 25-9: Water and Wastewater. Article 1: Utility Service. Division 3: Tap Permits. Section 25-9-99: Temporary Tap Permits for Community Gardens. See also Article 3: Water and Wastewater Recovery Fees. Division 4: Exemptions. Section 25-9-346: Exemption for Qualified Community Gardens (www.amlegal.com/nxt/gateway.dll/Texas/austin/thecodeofthecityofaustintexas?f=templates$fn=default.htm$3.0$vid=amlegal:austin_tx$anc=)	Authorizes the use of vacant lots in target revitalization areas and high-poverty area: community gardens. Exempts these lots fi platting requirements and water impact fe Authorizes temporary tap permits for gardening.
Allowances for the interim use of public land for urban agriculture			
Hartford, Conn.	Public land-use ordinance	Municipal Code. Chapter 26: Parks and Recreation. Article I: In General. Section 26-15: Municipal Garden Program (www.municode.com/resources/ gateway.asp?sid=7&pid=10895)	Establishes a program that authorizes the of surplus public lots for community garde
Utica, N.Y.	Public land-use ordinance	City Code. Part 1: Local Laws. Chapter 1-23: Streets, Sidewalks, and Public Places. Section 1-23-1: Community Garden Program (www.ecode360.com/?custid=ut2994)	Establishes a program that authorizes the of vacant public lots for community garder
Allowances for the use of public rights-of-way for curbside gardens			
Seattle	Internal policy	Department of Transportation Client Assistance Memo 2305: Gardening in Planting Strips (www.seattle.gov/transportation/stuse_garden.htm)	Clarifies that residents may plant raised-b gardens in the strip of the public right-of-w between the sidewalk and the curb after

(continued)

Public land disposition policies

City	Policy type	Description
Minneapolis	Public land disposition policy	Sets conditions for the use of vacant city land for community gardens; requires the purchaser to place a conservation easement on the community garden lot; allows the purchaser to construct accessory buildings on the lot for tools, equipment, and storage

Real Estate Disposition Policy (www.ci.minneapolis.mn.us/policies/disposition/disposition%20policy.doc)

Local food-purchasing policies

City	Policy type	Description
Toronto, Ont.	Local procurement policy	Sets a goal of purchasing 50 percent of all food for city facilities and operations from local sources (defined as within Ontario)
San Francisco	Executive directive	Directs all city departments and agencies to purchase local, sustainbly certified foods to the maximum extent possible

Local Food Procurement Policy (www.toronto.ca/legdocs/mmis/2008/cc/decisions/2008-10-29-cc25-dd.pdf)

Executive Directive 09-03: Healthy and Sustainable Food for San Francisco (www.sfgov3.org/ftp/uploadedfiles/sffood/policy_reports/MayorNewsomExecutiveDirectiveonHealthySustainableFood.pdf)

Standards for farmers markets that are city run or on public land

City	Policy type	Description
Chicago	Administrative ordinance	Outlines the rules and procedures for establishing city-run farmers markets
Philadelphia	Administrative ordinance	Specifies permitted locations for farmers markets and details operational and site standards
Sacramento, Calif.	Administrative ordinance	Authorizes the creation of a city-owned farmers market and outlines the basic requirements for sellers
San Francisco	Administrative ordinance	Outlines the requirements and operations standards for establishing a farmers market on city-owned property

Municipal Code. Title 4. Chapter 4-12: Farmers' Markets (www.amlegal.com/nxt/gateway.dll/Illinois/chicago_il/municipalcodeofchicago?f=templates$fn=default.htm$3.0$vid=amlegal:chicago_il)

Philadelphia Code. Title 9: Regulation of Businesses, Trades, and Professions. Chapter 9-200: Commercial Activities on Streets. Section 9-213: Farmers Markets (www.amlegal.com/nxt/gateway.dll/Pennsylvania/philadelphia_pa/thephiladelphiacode?f=templates$fn=default.htm$3.0$vid=amlegal:philadelphia_pa)

City Code. Title 5. Chapter 5.104: Producers' Market (www.qcode.us/codes/sacramento)

Administrative Code. Chapter 9A: Farmers' Market (http://library.municode.com/index.aspx?clientId=14131& stateId=5&stateName=California)

Stand-alone policies or directives

City	Policy type	Description
Seattle	Resolution	Gives specific assignments for a number of departments to help the city assess how it can better promote local food production and access. Directs the planning department to draft a code amendment to remove regulatory barriers to urban agriculture. Directs the Department of Neighborhoods to draft a Food Action Plan

Resolution 31019: Local Food Action Initiative (http://clerk.ci.seattle.wa.us/~scripts/nph-brs.exe?s1=&s2=&s3=31019&s4=&Sect4=AND&l=20&Sect2=THESON&Sect3=PLURON&Sect5=RESN1&Sect6=HITOFF&d=RES3&p=1&u=%2F~public%2Fresn1.htm&r=1&f=G)

(continued)

| San Francisco | Executive directive | Executive Directive 09-03: Healthy and Sustainable Food for San Francisco (www.sfgov3.org/ftp/uploadedfiles/sffood/policy_reports/MayorNewsomExecutiveDirectiveonHealthySustainableFood.pdf) | Forms a food policy council and tasks a number of city departments with specific actions to help promote access to healthy and sustainable food. Directs all agencies that control city land to determine which properties could be used for food production. Directs the planning department to integrate healthy and sustainable food goals into the general plan |
| Saanich, B.C. | Council policy | Council Policy 03/CW: Community Gardens (www.saanich.ca/living/pdf/communitygardenspolicy.pdf) | Expresses support for community gardens; establishes process for designation and development of parkland for community gardens |

Resources

APA Resources

Planning and Community Health Research Center
www.planning.org/nationalcenters/health

Specific information on food systems planning is at www.planning.org/nationalcenters/health/food.htm

"Community and Regional Food Planning" (*PAS Memo*, September 2007)
www.planning.org/pas/memo/index.htm

Enhancing Urban Food Systems (PAS Essential Info Packet 16)
www.planning.org/apastore/Search/Default.aspx?p=3853

Farmland Preservation (APA Education CD-ROM)
www.planning.org/apastore/Search/Default.aspx?p=3419

"Food Systems Planning" (*PAS QuickNotes*)
www.planning.org/pas/quicknotes/index.htm

Old Cities Green Cities: Communities Transform Unmanaged Land, by J. Baline Bonham Jr., Gerri Spilka, and Darl Rastorfer (PAS Report 506/507, 2002)
www.planning.org/apastore/Search/Default.aspx?p=2420

A Planners Guide to Community and Regional Food Planning, by Samina Raja, Branden Born, and Jessica Kozlowski Russell (PAS Report 554, 2008)
www.planning.org/apastore/search/Default.aspx?p=3886

Planning for Food Access (2009–present)
www.planning.org/research/foodaccess/index.htm

With funding from Healthy Eating Research, a National Program of the Robert Wood Johnson Foundation, the Planning and Community Health Research Center will identify and evaluate food access goals in comprehensive and sustainability plans across the country and manage the development of a report for policy makers that identifies best practices in planning for food access.

Planning Magazine, special issue on food (August/September 2009)
www.planning.org/planning/open/aug

Policy Guide on Agricultural Land Preservation
www.planning.org/policy/guides/adopted/agricultural.htm

Policy Guide on Community and Regional Food Planning
www.planning.org/policy/guides/adopted/food.htm

Regulating Temporary Summer Uses (PAS Essential Info Packet 9)
www.planning.org/apastore/Search/Default.aspx?p=3846

"Zoning for Public Markets and Street Vendors" (*Zoning Practice*, February 2009)
www.planning.org/zoningpractice/index.htm

"Zoning for Urban Agriculture" (*Zoning Practice*, March 2010)
www.planning.org/zoningpractice/index.htm

U.S. Environmental Protection Agency Resources

Urban Agriculture and Improving Local, Sustainable Food Systems
www.epa.gov/brownfields/urbanag/index.html

U.S. Department of Agriculture Resources

Community Food Projects
www.csrees.usda.gov/nea/food/in_focus/hunger_if_competitive.html

Farmers Market and Local Food Marketing
www.ams.usda.gov/AMSv1.0/FarmersMarkets

Food and Nutrition Service, Farm to School Initiative
www.fns.usda.gov/cnd/F2S/Default.htm

Food and Nutrition Service, FNS Farm to School Team
www.fns.usda.gov/cnd/F2S/f2stacticalteam.htm

Gardening with Children
http://healthymeals.nal.usda.gov/nal_display/index.php?info_center=14&tax_level=2&tax_subject=526&level3_id=0&level4_id=0&level5_id=0&topic_id=2112&&placement_default=0

People's Garden Initiative
www.usda.gov/wps/portal/usda/usdahome?navid=PEOPLES_GARDEN

Supplemental Nutrition Assistance Program, Farmers Market Voucher Program
www.fns.usda.gov/wic/FMNP

Urban Agriculture: An Abbreviated List of References and Resource Guide, September 2000
www.nal.usda.gov/afsic/AFSIC_pubs/urbanag.htm

Other Institutional Resources

American Community Gardening Association
www.communitygarden.org

ACGA is also the publisher of *Growing Communities: How to Build Community Through Community Gardening*, by Jeanette Abi-Nader, David Buckley, Kendall Dunnigan and Kristen Markley
www.communitygarden.org/acga-store.php#acgacategory

City Farmer (Canada's Office of Urban Agriculture)
www.cityfarmer.org

Community Food Security Coalition
www.foodsecurity.org

www.foodsecurity.org/FarmingCitytoFringe.pdf

www.foodsecurity.org/pubs.html#healthurbanag

Crossroads Research Center
Committee on Community Economic Development
www.crcworks.org/cfscced.html

Local Food as Economic Development fact sheet
www.crcworks.org/crcdocs/lfced.pdf

Food Security Learning Center
Community Economic Development resources
www.crcworks.org/cedresources.html

Community Gardens section
www.whyhunger.org/programs/fslc/topics/community-gardens.html

International Development Research Centre, Urban Agriculture for Sustainable Development
www.idrc.ca/in_focus_cities; http://publicwebsite.idrc.ca/EN/Pages/default.aspx

Cities Feeding People
www.idrc.ca/en/ev-8308-201-1-DO_TOPIC.html

Public Health Law and Policy

www.nplanonline.org/nplan/community-gardens

www.nplanonline.org/nplan/farmers-markets

Resource Centres on Urban Agriculture and Food Security

www.ruaf.org

Ryerson University

Carrot City Project

www.ryerson.ca/carrotcity

Other Publications

City Bountiful: A Century of Community Gardening in America

By Laura J. Lawson (University of California Press, 2005)

www.ucpress.edu/book.php?isbn=9780520243439

Progressive Planning

Issue 158, on food and planning (winter 2004)

www.plannersnetwork.org/publications/pdfs/2001-2004/PlannersNetwork_No158_062607.pdf

References

American Planning Association (APA). 2007. *Policy Guide on Community and Regional Food Planning.* Available at www.planning.org/policy/guides/adopted/food.htm.

———. 2010. "Food Systems Planning." *PAS QuickNotes,* no. 24. February.

Ashe, M., L. M. Feldstein, et al. 2007. "Local Venues for Change: Legal Strategies for Healthy Environments." *Journal of Law, Medicine & Ethics.* Spring, 138–47.

Austin (Tex.), City of. 2010. *City Code.* Section 25-2-863, "Urban Farms." Available at www .amlegal.com/austin_tx.

Baatz, Simon. 1985. *"Venerate the Plough": A History of the Philadelphia Society for Promoting Agriculture, 1785–1985.* Philadelphia: PSPA.

Babey, Susan, et al. 2008. *Designed for Disease: The Link Between Local Food Environments and Obesity and Diabetes.* Los Angeles: California Center for Public Health Advocacy, PolicyLink, and the UCLA Center for Health Policy Research. Available at www .healthpolicy.ucla.edu/pubs/Publication.aspx?pubID=250.

Bailkey, Martin. 2009. "An Update from New Orleans." *Urban Agriculture Magazine,* no. 22 (June). Available at www.ruaf.org/node/2071.

Balmer, Kevin, et al. 2005. *The Diggable City: Making Urban Agriculture a Planning Priority.* Portland, Ore.: Nohad A. Toulan School of Urban Studies and Planning, Portland State University. Available at http://diggablecity.org.

Baltimore, City of, Office of Sustainability. 2009. *Baltimore Sustainability Plan.* Available at www.baltimorecity.gov/LinkClick.aspx?fileticket=DtRcjL%2fIBcE%3d&tabid=128.

Behm, Don. 2009. "Growing Power Could Expand Food Programs in Deal with MMSD." *Milwaukee Journal Sentinel.* September 14. Available at www.jsonline.com/news/ milwaukee/59252472.html.

Bellows, Anne C., Katherine Brown, and Jac Smit. 2004. "Health Benefits of Urban Agriculture." Community Food Security Coalition's North American Initiative on Urban Agriculture. Available at www.foodsecurity.org/UAHealthArticle.pdf.

Bellows, B. C., R. Dufour, et al. 2003. "Bringing Local Food to Local Institutions: A Resource Guide for Farm-to-School and Farm-to-Institution Programs." National Center for Appropriate Technology ATTRA Resource Series. Available at http://attra.ncat.org/ attra-pub/PDF/farmtoschool.pdf.

Boston, City of. 2009. *Zoning Code.* Chapter 33, "Open Space Subdistricts." Available at www.amlegal.com/boston_ma.

Breaking Through Concrete. 2010a. "Chicagoans Get New Roots and Second Chances from Growing Home Farm." *Grist.* July 16. Available at www.grist.org/article/food-Chicagoans-get-new-roots-and-second-chances-from-Growing-Home-farm-.

———. 2010b. "Philly's Greensgrow Farm: An Unconventional Hybrid That Works." *Grist.* July 5. Available at www.grist.org/article/food-Phillys-Greensgrow-farm-an-unconventional-hybrid-that-works.

Brown, K. H., and A. Carter. 2003. "Urban Agriculture and Community Food Security in the United States: Farming from the City Center to the Urban Fringe." Community Food Security Coalition, North American Urban Agriculture Committee. Available at www.foodsecurity.org/PrimerCFSCUAC.pdf.

Buckley, C. 2009. "Two Acres of Hope for Recovering Addicts." *New York Times.* August 14. Available at www.nytimes.com/2009/08/16/nyregion/16farm.html?_r=1.

Caton Campbell, Marcia. 2004. "Building a Common Table: The Role for Planning in Community Food Systems." *Journal of Planning Education and Research* 23(4): 341–55.

Caton Campbell, Marcia, and Danielle A. Salus. 2003. "Community and Conservation Land Trusts as Unlikely Partners? The Case of Troy Gardens, Madison, Wisconsin." *Land Use Policy* 20(2): 169–80.

Center for Civic Partnerships. 2003. *From Organizational Practices to Public Policies: Local Strategies to Increase Healthy Eating & Physical Activity.* Sacramento, Calif.: Public Health Institute. Available at www.phi.org/pdf-library/ccppolicybrief.pdf.

Chattanooga (Tenn.), City of. 2009. *Zoning Regulations.* Section 1600, "A-1 Urban Agricultural Zone." Available at www.chattanooga.gov/City_Council/110_Code.asp.

Cheadle, A., B. Psaty, et al. 1993. "Can Measures of the Grocery Store Environment Be Used to Track Community-Level Dietary Changes?" *Preventive Medicine* 22(3): 361–72.

Chicago, City of. 2010. *Municipal Code.* Section 7-28-710, "Dumping Prohibited"; Section 7-28-715, "Composting Standards." Available at www.amlegal.com/library/il/chicago.shtml.

Chicago, City of, Department of the Environment. n.d. "Green Job Factsheet." Available at www.cityofchicago.org/content/dam/city/depts/doe/general/PressRelease PDFs/GreenJobsContactSheet.pdf.

Cleveland, City of. 2010. *Codified Ordinances.* Section 347.02, "Restrictions on the Keeping of Farm Animals and Bees." Available at http://caselaw.lp.findlaw.com/clevelandcodes.

Cleveland, City of, City Planning Commission. 2008. "Re-Imagining a More Sustainable Cleveland: Citywide Strategies for Reuse of Vacant Land." Adopted December 19. Available at http://neighborhoodprogress.org/uploaded_pics/reimagining_final_screen-res_file_1236290773_file_1241529460.pdf.

Cleveland–Cuyahoga County Food Policy Coalition. n.d. "Policy Brief: Water Access for Urban Agriculture and Greening Projects." Draft. Available via www.cccfoodpolicy.org.

Cleveland Urban Design Collaborative (CUDC), Kent State University. 2008. *Re-Imagining a More Sustainable Cleveland: Citywide Strategies for Reuse of Vacant Land.* Available at www.cudc.kent.edu/shrink/Images/reimagining_final_screen-res.pdf.

Community Food System Coalition. 2007. "The North American Urban and Peri-Urban Agriculture Alliance." Available at www.foodsecurity.org/NAUPAA_description_Nov_2007.pdf.

———. 2010. "What Is Community Food Security?" Available at www.foodsecurity.org/views_cfs_faq.html.

Crescent City Farmers Market. n.d. "New Orleans Market History." Available at www.crescentcityfarmersmarket.org/index.php?page=new-orleans-market-history.

Daniels, Tom. 2008. "Taking the Initiative: Why Cities Are Greening Now." Pp. 259–78 in *Growing Greener Cities: Urban Sustainability in the Twenty-First Century,* ed. Eugenie L. Birch and Susan M. Wachter. Philadelphia: University of Pennsylvania Press.

Davis, John Emmeus, ed. 2010. *The Community Land Trust Reader.* Part Five: Beyond Housing. Cambridge, Mass.: Lincoln Institute of Land Policy.

de la Salle, Janine, and Mark Holland, eds. 2010. *Agricultural Urbanism: Handbook for Building Sustainable Food Systems in 21st Century Cities.* Winnipeg, Manitoba: Green Frigate Books.

de Zeeuw, H., M. Dubbeling, et al. 2007. "Key Issues and Courses of Action for Municipal Policy Making on Urban Agriculture." *RUAF Working Paper Series.* Leusden, The Netherlands: Resource Centres on Urban Agriculture and Food Security. Available at www.ruaf.org/node/2292.

Delaware Valley Regional Planning Commission (DVRPC). 2010. *Greater Philadelphia Food Systems Study.* Publication 09066A. Available at www.dvrpc.org/asp/pubs/publicationabstract.asp?pub_id=09066A.

———. 2009. *Connections: The Regional Plan for a Sustainable Future.* Available at www.dvrpc.org/Connections/.

Denver, City of. 2010. *Zoning Code.* Available at http://denvergov.org/cpd/zoning/DenverZoningCode/tabid/432507/Default.aspx.

Donofrio, Gregory. 2008. "Feeding the City." *Gastronomica* 7(4): 30–41.

Duany Plater-Zyberk and Company (DPZ). 2008. "SmartCode Modules." Available at www.smartcodecentral.org.

Dubbeling, Marielle, Marcia Caton Campbell, Femke Hoekstra, and René van Veehuizen. 2009. "Building Resilient Cities." *Urban Agriculture Magazine,* no. 22 (June): 3–11.

Dubbeling, M., and G. Merzthal. 2006. "Sustaining Urban Agriculture Requires the Involvement of Multiple Stakeholders." Chap. 2 in Veenhuizen 2006.

Eggler, B. 2010. "New Orleans Master Plan Approved by City Council." *The* (New Orleans) *Times Picayune.* August 12. Available at www.nola.com/politics/index.ssf/2010/08/new_orleans_master_plan_approv.html.

Escondido (Calif.), City of. n.d. "Adopt-a-Lot Program." Available at www.escondido.org/adopt-a-lot-program.aspx.

Estes, Emily, Megan R. Carter-Thomas, and Daniel J. Brabander. 2010. "Deposition of Particulate Matter as a Mechanism for Trace Metal Contamination of Urban Gardens." Paper No. 141-12, 2010 Geological Society of America Annual Meeting, Denver, 31 October–3 November. Available at http://gsa.confex.com/gsa/2010AM/finalprogram/abstract_182493.htm.

Federal Reserve Bank of Atlanta. 2009. "Land Banking as a Foreclosure Mitigation Strategy." September. Available at www.frbatlanta.org/podcasts/transcripts/foreclosureresponse/09Sep_land_banking.cfm.

Feldstein, L. 2007. "Linking Land Use Planning and the Food Environment." *Smart Growth Online.* January.

Flournoy, R., and S. Treuhaft. 2005. *Healthy Food, Healthy Communities: Improving Access and Opportunities Through Food Retailing.* Oakland, Calif.: PolicyLink. Available at www.policylink.org/atf/cf/%7B97C6D565-BB43-406D-A6D5-ECA3BBF35AF0%7D/HEALTHYFOOD.pdf.

Forum of Research Connections (FORC), Herb Barbolet, Vijay Cuddeford, Fern Jeffries, Holly Korstad, Susan Kurbis, Sandra Mark, Christiana Miewald, and Frank Moreland. n.d. *Vancouver Food System Assessment.* Available at http://vancouver.ca/commsvcs/socialplanning/initiatives/foodpolicy/tools/pdf/vanfoodassessrpt.pdf.

Goody Clancy. 2010. *Plan for the 21st Century: New Orleans 2030.* Vol. 2, *Strategies and Actions.* Available at www.nolamasterplan.org/documentsandrresources.asp#C12.

Greater London Authority. 2010. *Cultivating the Capital: Food Growing and the Planning System in London.* Available at www.london.gov.uk/who-runs-london/the-london-assembly/publications/housing-planning/cultivating-capital-food-growing-and-planning-system-london.

Greensboro (N.C.), City of. 2010. *Code of Ordinances*. Section 30-4-5.1, "Permitted Uses." Available at http://library1.municode.com/default-test/home.htm?infobase=10736&doc_action=whatsnew.

Groc, I. 2008. "Growers' Gamble: Are Cities Missing a Bet by Giving Short Shrift to Farmers' Markets?" *Planning*. March: 34–37.

Gustafson, A., D. Cavallo, et al. 2007. "Linking Homegrown and Locally Produced Fruits and Vegetables to Improving Access and Intake in Communities through Policy and Environmental Change." *Journal of the American Dietetic Association* 107(4): 584–85.

Harmon, A. 2003. "Farm to School: An Introduction for Food Service Professionals, Food Educators, Parents, and Community Leaders." National Farm to School Program, Center for Food and Justice, Urban and Environmental Policy Institute.

Harmon, A. H., and B. L. Gerald. 2007. "Position of the American Dietetic Association: Food and Nutrition Professionals Can Implement Practices to Conserve Natural Resources and Support Ecological Sustainability." *Journal of the American Dietetic Association* 107(6): 1033–43.

Harper, A., A. Shattuck, E. Holt-Gimenez, A. Alkon, and F. Lambrick. 2009. *Food Policy Councils: Lessons Learned*. Institute for Food and Development Policy. Available at www.foodfirst.org/en/foodpolicycouncils-lessons.

Hartford (Conn.), City of. 2010. *Municipal Code*. Section 26-15, "Municipal Garden Program." Available at http://library6.municode.com/default-test/home.htm?infobase=10895&doc_action=whatsnew.

Haslett-Marroquin, Reginaldo. 2009. "Marketing and Distribution Partnership for 'Pollo de Campo' Gets Moving!" June 10. Available at http://www.ruralec.com/archives/537.

HB Lanarc. n.d. "Agricultural Urbanism." Available at www.agriculturalurbanism.com.

Hedden, W. P. 1929. *How Great Cities Are Fed*. Boston: D. C. Heath.

Hersh, R., D. Morley, J. Schwab, and L. Solitare. 2010. *Creating Community-Based Redevelopment Strategies*. Chicago: American Planning Association. Available at www.planning.org/research/brownfields/index.htm.

Hollander, Justin B., Niall G. Kirkwood, and Julia L. Gold. 2010. *Principles of Brownfield Regeneration: Cleanup, Design, and Reuse of Derelict Land*. Washington, D.C.: Island Press.

Horowitz, C., K. Colson, et al. 2004. "Barriers to Buying Healthy Foods for People with Diabetes: Evidence of Environmental Disparities." *American Journal of Public Health* 94(9): 1549–54.

Hynes, H. Patricia. 1996. *A Patch of Eden: America's Inner-City Gardeners*. White River Junction, Vt.: Chelsea Green.

Innes, J. E., and D. E. Booher. 1999. "Consensus Building and Complex Adaptive Systems: A Framework for Evaluating Collaborative Planning." *Journal of the American Planning Association* 65(4): 412–23.

Institute for Agriculture and Trade Policy (IATP). n.d. "The Minneapolis Mini Farmers Market Project." Available at www.iatp.org/iatp/publications.cfm?refid=103490%20.

Johnson, Lorraine. 2010. *City Farmer: Adventures in Urban Food Growing*. Vancouver, B.C.: Greystone Books.

Kansas City Center for Urban Agriculture (KCCUA). n.d. "Growing Good Food in Kansas City Neighborhoods: A Guide to Urban Agriculture Codes in KCMO." Available at www.kccua.org/KCMO%20Urban%20Ag%20Codes%20Guide%20-%20booklet.pdf.

Kaufman, J., and M. Bailkey. 2000. "Farming Inside Cities: Entrepreneurial Urban Agriculture in the United States." Lincoln Institute of Land Policy Working Paper. Available at www.lincolninst.edu/pubs/dl/95_KaufmanBaikey00.pdf.

Kaufman, Jerome L., and Martin Bailkey. 2004. "Farming Inside Cities through Entrepreneurial Urban Agriculture." Pp. 177–99 in *Recycling the City: The Use and Reuse of Urban Land*, ed. Rosalind Greenstein and Yesim Sungu-Eryilmaz. Cambridge, Mass.: Lincoln Institute of Land Policy.

Kelly, E. D., and B. Becker. 2000. *Community Planning: An Introduction to the Comprehensive Plan.* Washington, D.C.: Island Press.

Lawson, Laura J. 2005. *City Bountiful: A Century of Community Gardening in America.* Berkeley: University of California Press.

Levi, J., S. Vintner, et al. 2010. *F as in Fat: How Obesity Policies Are Failing in America.* Washington D.C.: Trust for America's Health. Available at http://healthyamericans.org/reports/obesity2010.

Lyson, Thomas A. 2005. "Civic Agriculture and Community Problem Solving." *Culture and Agriculture* 27(2): 92–98.

McCann, B. 2006. *Community Design for Healthy Eating: How Land Use and Transportation Solutions Can Help.* Princeton, N.J.: Robert Wood Johnson Foundation. Available at www.rwjf.org/files/publications/other/communitydesignhealthyeating.pdf.

McClintock, Nathan, and Jenny Cooper. 2009. "Cultivating the Commons: An Assessment of the Potential for Urban Agriculture on Oakland's Public Land." Available at http://urbanfood.org/docs/Cultivating_the_Commons2010_LoRes.pdf.

McCormack, Lacey Arneson, Melissa Nelson Laska, Nicole I. Larson, and Mary Story. 2010. "Review of the Nutritional Implications of Farmers' Markets and Community Gardens: A Call for Evaluation and Research Efforts." *Journal of the American Dietetic Association* 110(3): 399–408.

Madison (Wis.), City of. 2006. City of Madison Comprehensive Plan. Available at www.cityofmadison.com/planning/comp.

Mallach, A. 2006. *Bringing Buildings Back: From Abandoned Properties to Community Assets.* Montclair, N.J.: National Housing Institute.

Mendes, W. 2008. "Implementing Social and Environmental Policies in Cities: The Case of Food Policy in Vancouver, Canada." *International Journal of Urban and Regional Research* 32: 942–67.

Michigan Municipal League. 2010. "Flint-Area Couple Grow Urban Agriculture Trend." Available at www.mml.org/newsroom/press_releases/2010_07_15.html.

Michigan State University. 2010. "Urban Farms Could Provide a Majority of Produce for Detroiters." November 16. Available at http://news.msu.edu/story/8600.

Minneapolis, City of. 2010. *Code of Ordinances.* Chapter 527, "Planned Unit Development." Available at http://library1.municode.com/default-test/home.htm?infobase=11490&doc_action=whatsnew.

Minneapolis, City of, Community Planning and Economic Development, Planning Division. 2010a. "City Parcels Available for 2010 Community Garden Pilot: Homegrown Minneapolis." December 1. Available from www.ci.minneapolis.mn.us/dhfs/homegrown-home.asp.

———. 2010b. "Urban Agriculture Policy Plan." Available at www.ci.minneapolis.mn.us/cped/urban_ag_plan.asp.

Minneapolis Department of Health and Family Support (MDHFS). 2009. *Homegrown Minneapolis: Final Report Presented to the Health, Energy and Environment Committee of the Minneapolis City Council.* June 15. Available at www.ci.minneapolis.mn.us/dhfs/hgfinalrec.pdf.

Mooallem, J. 2008. "Guerilla Gardening." *New York Times Magazine.* June 8, 76–82. Available at www.nytimes.com/2008/06/08/magazine/08guerrilla-t.html.

Mougeot, L. J. A. 1999. "Urban Agriculture: Definition, Presence, Potential and Risks, Main Policy Challenges." *International Workshop on Growing Cities Growing Food: Urban Agriculture on the Policy Agenda.* Havana, Cuba: IDRC. Available at www.idrc.ca/en/ev-2571-201-1-DO_TOPIC.html.

Mubvami, T., and S. Mushamba. 2006. "Integration of Agriculture in Urban Land Use Planning." Chap. 3 in Veenhuizen 2006.

Mukherji, Nina, and Alfonso Morales. 2010. "Zoning for Urban Agriculture." *Zoning Practice*. March.

Myers, Albert Cook. 1912. *Narratives of Early Pennsylvania, West New Jersey and Delaware, 1630–1707*. New York: Scribner's Sons.

Nasr, Joseph, Rod MacRae, and James Kuhns. 2010. *Scaling up Urban Agriculture in Toronto: Building the Infrastructure*. Toronto: Metcalf Foundation. Available at www.metcalffoundation .com/downloads/Metcalf_Food_Solutions_Scaling_Up_Urban_Agriculture_in_Toronto .pdf.

National Policy and Legal Analysis Network to Prevent Childhood Obesity (NPLAN). 2009. *Establishing Land Use Protections for Farmers' Markets*. December. Available at www .nplanonline.org/nplan/products/establishing-land-use-protections-farmers-markets.

National Research Council. 2010. *Toward Sustainable Agriculture Systems in the 21st Century*. Washington, D.C.: National Academies Press. Available at www.nap.edu/catalog .php?record_id=12832#toc.

New Orleans, City of. 2008. *Comprehensive Zoning Ordinance*. Section 4.1.3, "RS-1 Single-Family Residential District – Permitted Uses." Section 4.1.4, "RS-1 Single-Family Residential District – Accessory Uses." Available at http://library.municode.com/ index.aspx?clientId=16306&stateId=18&stateName=Louisiana.

New Orleans Food Policy Advisory Committee (FPAC). 2008. *Building Healthy Communities: Expanding Access to Fresh Food Retail*. March. Available at www.sph.tulane.edu/PRC/ Files/FPAC%20Report%20Final.pdf.

Newman, Peter, Timothy Beatley, and Heather Boyer. 2009. *Resilient Cities: Responding to Peak Oil and Climate Change*. Washington, D.C.: Island Press.

Nordahl, Darrin. 2009. *Public Produce: The New Urban Agriculture*. Washington, D.C.: Island Press.

Northridge, Mary E., Elliott D. Sclar, and Padmini Biswas. 2003. "Sorting Out the Connections Between the Built Environment and Health: A Conceptual Framework for Navigating Pathways and Planning Healthy Cities." *Journal of Urban Health* 80(4): 556–68.

Ohio State University Extension. n.d. "Market Gardening and Urban Farming." Available at http://cuyahoga.osu.edu/topics/agriculture-and-natural-resources/market-gardening-and-urban-farming-1.

Olopade, Dayo. 2009. "Green Shoots in New Orleans." *The Nation*. September 21, 21–26.

Philadelphia, City of. 2010a. *Philadelphia Code*. Section 9-213, "Farmers' Markets." Available at www.amlegal.com/library/pa/philadelphia.shtml.

———. 2010b. *Zoning Code*. Module 2, "Zoning Districts and Uses" (proposed). Available at www.zoningmatters.org.

Plyer, Allison, Elaine Ortiz, and Kathryn L. S. Pettit. 2010. "Optimizing Blight Strategies: Deploying Limited Resources in Different Neighborhood Housing Markets." New Orleans: Greater New Orleans Community Data Center; available at www.gnocdc .org/OptimizingBlightStrategies/index.html.

Pollack, C. 2010. "Learn How Drip Irrigation Can Boost on-Farm Profits at July 15 Workshop." Ohio State University Extension, June 17. Available at www.ag.ohio-state .edu/~news/story.php?id=5765.

Portland (Ore.), City of, Bureau of Planning and Sustainability. 2010. "Urban Growth Bounty 2010." Available at www.portlandonline.com/bps/index.cfm?c=50648&.

Pothukuchi, K., H. Joseph, H. Burton, and A. Fisher. 2002. *What's Cooking in Your Food System: A Guide to Community Food Assessment*. Community Food Security Coalition. Available at http://foodsecurity.org/pub/whats_cooking.pdf.

Pothukuchi, Kameshwari, and Jerome L. Kaufman. 1999. "Placing the Food System on the Urban Agenda: The Role of Municipal Institutions in Food Systems Planning." *Agriculture and Human Values* 16(2): 213–24.

————. 2000. "The Food System: A Stranger to Urban Planning." *Journal of the American Planning Association* 66(2): 113–24.

Powell, L. M., F. J. Chaloupka, et al. 2007. "The Availability of Fast-Food and Full-Service Restaurants in the United States: Associations with Neighborhood Characteristics." *American Journal of Preventive Medicine* 33(4S): S240–45.

Powell, L. M., S. Slater, et al. 2007. "Food Store Availability and Neighborhood Characteristics in the United States." *Preventive Medicine* 44: 189–95.

Raja, Samina, Branden Born, and Jessica Kozlowski Russell. 2008. *A Planners Guide to Community and Regional Food Planning.* Planning Advisory Service Report no. 554. Chicago: American Planning Association.

Rauzon, S., M. Wang, N. Studer, and P. Crawford. 2010. *An Evaluation of the School Lunch Initiative, Final Report.* Berekley, Calif.: Dr. Robert C. and Veronica Atkins Center for Weight and Health, College of Natural Resources and School of Public Health, University of California, Berkeley. Available at www.ecoliteracy.org/downloads/school-lunch-initiative-evaluation.

Region of Waterloo Public Health. 2007. *A Healthy Community Food System Plan for Waterloo Region.* Available at http://chd.region.waterloo.on.ca/web/health.nsf/8f9c046037662cd985256af000711418/54ED787F44ACA44C852571410056AEB0/$file/FoodSystem_Plan.pdf?openelement.

Reinhardt, C. n.d. "Farming in the 1940s: Victory Gardens." York, Neb.: Wessels Living History Farm. Available at www.livinghistoryfarm.org/farminginthe40s/crops_02.html.

Richmond (Calif.), City of. 2009. *Richmond General Plan.* Element 11, Community Health and Wellness. Available at www.cityofrichmondgeneralplan.org/docManager/1000000797/11%20Health.pdf.

Roberts, Wayne. 2008. *The No-Nonsense Guide to World Food.* Toronto: New Internationalist Publications.

Rolls, B. J., J. A. Ello-Martin, et al. 2004. "What Can Intervention Studies Tell Us About the Relationship Between Fruit and Vegetable Consumption and Weight Management?" *Nutrition Review* 62(1): 1–17.

Rose, Donald, J. Nicholas Bodor, Chris M. Swalm, Janet C. Rice, Thomas A. Farley, and Paul L. Hutchinson. 2009. "Deserts in New Orleans? Illustrations of Urban Food Access and Implications for Policy." February. Available at www.npc.umich.edu/news/events/food-access/rose_et_al.pdf.

Rosenberg, Greg. 2010. "Troy Gardens: The Accidental Ecovillage." Pp. 420–31 in *The Community Land Trust Reader*, ed. John Emmeus Davis. Cambridge, Mass.: Lincoln Institute of Land Policy.

Rosenzweig, Roy, and Elizabeth Blackmar. 1992. *The Park and the People: A History of Central Park.* Ithaca, N.Y.: Cornell University Press.

Sacramento (Calif.), City of. 2010. *City Code.* Chapter 5.104, "Producers' Market." Available at www.qcode.us/codes/sacramento.

Safety Harbor (Fla.), City of. 2008. *Comprehensive Zoning and Land Development Code.* Section 271.00(32), "Definitions: Community Garden." Available at http://library8.municode.com/default-test/home.htm?infobase=11301&doc_action=whatsnew.

Sallis, J. F., and K. Glanz. 2006. "The Role of the Built Environments in Physical Activity, Eating, and Obesity in Childhood." *The Future of Children* 16(1): 89–108.

Sallis, J. F., P. Nader, et al. 1986. "San Diego Surveyed for Heart-Healthy Foods and Exercise Facilities." *Public Health Report* 101(2): 216–19.

San Antonio (Tex.), City of. 2010. *Code of Ordinances.* Section 5-52, "Keeping of Bovines, Equines, Sheep, and Goats"; Section 5-109, "Animal Limits; Excess Animal Permit"; Section 5-114, "Livestock Permits." Available at http://library.municode.com/index.aspx?clientId=11508&stateId=43&stateName=Texas.

San Francisco, City and County of, Department of the Environment. n.d. "Composting." Available at www.sfenvironment.org/our_programs/topics.html?ssi=3&ti=6.

San Francisco, City and County of, Office of the Mayor. 2009. *Executive Directive 09-03*. Available at www.sfgov3.org/ftp/uploadedfiles/sffood/policy_reports/MayorNewsomExecutive DirectiveonHealthySustainableFood.pdf.

Sauder, Robert A. 1981. "The Origin and Spread of the Public Market System in New Orleans." *Louisiana History: The Journal of the Louisiana Historical Association* 22(3): 281–97.

Schukoske, J. E. 2000. "Community Development Through Gardening: State and Local Policies Transforming Urban Open Space." *Legislation and Public Policy* 3: 351–92.

Shayler, Hannah, Murray McBride, and Ellen Harrison. 2009. "Sources and Impacts of Contaminants in Soils." Ithaca, N. Y.: Cornell Waste Management Institute. Available at http://cwmi.css.cornell.edu/sourcesandimpacts.pdf.

Seattle, City of. 2005 / 2007. *Seattle's Comprehensive Plan, Urban Village Element*. Available at www.seattle.gov/DPD/static/Urban%20Village%20element_LatestReleased_DPDP016169.pdf.

———. 2010a. *Municipal Code*. Section 23.42.052, "Keeping of Animals." Available at http://clerk.ci.seattle.wa.us/~public/code1.htm.

———. 2010b. *Ordinance 123378*. Available at www.seattle.gov/DPD/Planning/Urban Agriculture and http://clerk.ci.seattle.wa.us/~archives/Ordinances/Ord_123378.pdf.

Seattle, City of, Department of Transportation (DOT). 2009. "Gardening in Planting Strips." Client Assistance Memo 2305. Available at www.seattle.gov/transportation/stuse_garden.htm.

Smit, J., and M. Bailkey. 2006. "Urban Agriculture and the Building of Communities." Chap. 6 in *Cities Farming for the Future: Urban Agriculture for Green and Productive Cities*. R. v. Veenhuizen, RUAF Foundation, IDRC and IIRR.

South Central Farm. 2005. "What We Are About?" October 3. Available at www.southcentralfarmers.com/index.php?option=com_content&task=view&id=12&Itemid=9.

Stair, Peter, Heather Wooten, and Matt Raimi. 2008. *How to Create and Implement Healthy General Plans: The General Plan as a Tool for Change*. Oakland, Calif.: Public Health Law & Policy and Raimi + Associates. Available at www.phlpnet.org/system/files/Healthy_General_Plans_Toolkit-WEB.pdf.

Stevenson, G. W., Kathryn Ruhf, Sharon Lezberg, and Kate Clancy. 2007. "Warrior, Builder, and Weaver Work: Strategies for Changing the Food System." Pp. 33-62 in *Remaking the North American Food System: Strategies for Sustainability*, ed. C. Clare Hinrichs and Thomas A. Lyson. Lincoln: University of Nebraska Press.

Sullivan, Dan. 2006. "Small Is Beautiful and Profitable." *New Farm*. June 8. Available at http://newfarm.rodaleinstitute.org/features/2006/0606/somertontanks/sullivan.shtml.

Teig, Ellen, Joy Amulya, Lisa Bardwell, Michael Buchenau, Julie A. Marshall, and Jill S. Litt. 2009. "Collective Efficacy in Denver, Colorado: Strengthening Neighborhoods and Health through Community Gardens." *Health & Place* 15(4): 1115–22.

Tidball, Keith G., and Marianne E. Krasny. 2007. "From Risk to Resilience: What Role for Community Greening and Civic Ecology in Cities?" Pp. 149–64 in *Social Learning Towards a More Sustainable World*, ed. Arjen Wals. Wageningen, The Netherlands: Wageningen Academic Press.

Tixier, P., and H. d. Bon. 2006. "Urban Horticulture." Chap. 11 in Veenhuizen 2006.

Toledo (Ohio), City of. 2010. *Municipal Code*. Section 1116.0202, "Agriculture." Available at www.amlegal.com/toledo_oh.

Toronto (Ontario), City of. 2007. "Change Is in the Air: Climate Change, Clean Air and Sustainable Energy Action Plan: Moving from Framework to Action. Phase 1." June. Available at www.toronto.ca/changeisintheair/pdf/clean_air_action_plan.pdf.

Toronto (Ontario), City of, Medical Officer of Health. 2010. "Toronto Food Strategy: Culti-vating Food Connections." May 20. Available at www.toronto.ca/legdocs/mmis/2010/hl/bgrd/backgroundfile-30482.pdf.

Tranel, M. and L. B. Handlin, Jr. 2006. "Metromorphosis: Documenting Change." *Journal of Urban Affairs* 28: 151–67.

Tropp, D., and S. Olowolayemo. 2000. "How Local Farmers and School Food Service Buy-ers Are Building Alliances: Lessons Learned from the USDA Small Farm/School Meals Workshop." Washington D.C.: USDA, Agricultural Marketing Service.

Turner, Allison H. 2009. *Urban Agriculture and Soil Contamination: An Introduction to Urban Gardening.* Louisville, Ky.: University of Louisville Center for Environmental Policy and Management. Available at http://cepm.louisville.edu/Pubs_WPapers/practice guides/PG25.pdf.

U.S. Department of Agriculture, Agricultural Marketing Service (USDA AMS). 2010. "Farmers Market Growth: 1994–2010." Available at www.ams.usda.gov/AMSv1.0/ams.fetchTemplateData.do?template=TemplateS&navID=WholesaleandFarmersMar kets&leftNav=WholesaleandFarmersMarkets&page=WFMFarmersMarketGrowth& description=Farmers%20Market%20Growth&acct=frmrdirmkt.

U.S. Department of Agriculture, Economic Research Service (USDA ERS). 2009. "Briefing Rooms—Farm and Commodity Policy: Questions and Answers." Available at www .ers.usda.gov/briefing/farmpolicy/cost2008bill.htm.

U.S. Department of Agriculture, National Institute of Food and Agriculture (USDA NIFA). 2008. "About Us: Cooperative Extension System Offices." Available at www.csrees .usda.gov/Extension.

U.S. Department of Health and Human Services, Assistant Secretary for Planning and Evaluation (U.S. HHS–ASPE). 2003. *Prevention Makes Common "Cents."* Available at http://aspe.hhs.gov/health/prevention.

U.S. Department of Health and Human Services (U.S. HHS) and U.S. Department of Agri-culture (USDA). 2005. *Dietary Guidelines for Americans, 2005.* Washington D.C.: U.S. GPO. Available at www.health.gov/dietaryguidelines/dga2005/document/default.htm.

U.S. Environmental Protection Agency (U.S. EPA). 2009. "How Does Your Garden Grow? Brownfields Redevelopment and Local Agriculture." EPA-560-F-09-024.

———. 2010a. "HUD-DOT-EPA Interagency Partnership for Sustainable Communities." Available at www.epa.gov/dced/partnership/index.html.

———. 2010b. "Safe Gardening, Safe Play, and a Safe Home." Iron King Mine and Humboldt Smelter Superfund Site Fact Sheet. Available at http://yosemite.epa.gov/r9/sfund/r9sfdocw.nsf/3dc283e6c5d6056f88257426007417a2/e03731cfc03cb5fa882576ea006f514c/$FILE/Safe%20Gardening%20in%20Dewey-Humboldt.pdf.

———. n.d. "Environmental Justice Showcase Communities." Available at www.epa.gov/compliance/ej/grants/ej-showcase.html.

Vancouver (British Columbia), City of. 2005. "Policy Report, Social Development: Hobby Beekeeping (Urban Apiculture) in Vancouver." July 5. Available at http://vancouver .ca/ctyclerk/cclerk/20050721/documents/pe3.pdf.

———. 2009. *Urban Agriculture Guidelines for the Private Realm.* Available at http://vancouver .ca/ctyclerk/cclerk/20090120/documents/p2.pdf.

———. n.d. "Southeast False Creek." Available at http://vancouver.ca/commsvcs/southeast/index.htm.

Vancouver (British Columbia), City of, Animal Control. n.d. "Steps to Keeping Backyard Hens: Regulations." Available at http://vancouver.ca/commsvcs/LICANDINSP/animalcontrol/chicken/index.htm.

Vancouver (British Columbia), City of, Community Services, Social Planning. 2009. "Food Policy—Food Policy in Vancouver: A Short History." Available at http://vancouver.ca/commsvcs/socialplanning/initiatives/foodpolicy/policy/history.htm.

———. 2010. "Food Policy—Community Gardens & the 2010 Challenge." Available at http://vancouver.ca/commsvcs/socialplanning/initiatives/foodpolicy/projects/2010gardens.htm.

Vancouver (British Columbia), City of, Vancouver Food Policy Council. 2009. Meeting minutes, February 11. Available at http://vancouver.ca/COMMSVCS/socialplanning/initiatives/foodpolicy/policy/pdf/090211_minutes.pdf.

Vancouver (British Columbia), City of, Vancouver Food Policy Task Force. 2003. "Action Plan for Creating a Just and Sustainable Food System for the City of Vancouver." Policy report. December 9. Available at http://vancouver.ca/ctyclerk/cclerk/20031209/rr1.htm.

Veenhuizen, R. v. 2006. *Cities Farming for the Future: Urban Agriculture for Green and Productive Cities.* Silang, Cavite (Philippines): International Institute of Rural Reconstruction and ETC Urban Agriculture. Available at www.idrc.ca/openebooks/216-3.

Viljoen, A., ed. 2005. *Continuous Productive Open Spaces: Designing Urban Agriculture for Sustainable Cities.* Oxford: Architectural Press.

Vitiello, Domenic. 2008. "Growing Edible Cities." Pp. 259–78 in *Growing Greener Cities: Urban Sustainability in the Twenty-First Century*, ed. Eugenie L. Birch and Susan M. Wachter. Philadelphia: University of Pennsylvania Press.

———. Forthcoming. "Planning the Food Secure City: Philadelphia Agriculture, Retrospect and Prospect." In *Nature's Entrepot: Philadelphia's Urban Sphere and Its Environmental Thresholds*, ed. Michael Chiarrappa and Brian Black. Pittsburgh: University of Pittsburgh Press.

Vitiello, Domenic, and Michael Nairn. 2009. "Community Gardening in Philadelphia: 2008 Harvest Report." Penn Planning and Urban Studies, University of Pennsylvania. Available at https://sites.google.com/site/urbanagriculturephiladelphia/home.

Voicu, I., and Been, V. 2008. "The Effect of Community Gardens on Neighboring Property Values." *Real Estate Economics* 36: 241–83. Available at http://furmancenter.org/files/publications/The_Effect_of_Community_Gardens.pdf.

Ward, Andrea. 2010. "Asphalt Garden." McGraw Hill Construction Continuing Education Center, May. Available at http://continuingeducation.construction.com/article.php?L=5&C=679&P=2.

Warner, Sam Bass. 1987. *To Dwell Is to Garden: A History of Boston's Community Gardens.* Boston: Northeastern University Press.

Whitesall, Amy. 2007. "Detroit—From Plot to Plate: SE Michigan's Urban Gardens." *Metromode.* October 18. Available at www.metromodemedia.com/features/UrbanGardens0041.aspx?referrerID=fbce1ca4-4d52-4248-90c8-1d38b1b5f85d.

World Food Summit. 1996. "Rome Declaration on World Food Security." November 13–17. Available at www.fao.org/docrep/003/w3613e/w3613e00.HTM.

APA American Planning Association

Making Great Communities Happen

The American Planning Association provides leadership in the development of vital communities by advocating excellence in community planning, promoting education and citizen empowerment, and providing the tools and support necessary to effect positive change.

518/519. Ecological Riverfront Design. Betsy Otto, Kathleen McCormick, and Michael Leccese. March 2004. 177pp.

520. Urban Containment in the United States. Arthur C. Nelson and Casey J. Dawkins. March 2004. 130pp.

521/522. A Planners Dictionary. Edited by Michael Davidson and Fay Dolnick. April 2004. 460pp.

523/524. Crossroads, Hamlet, Village, Town (revised edition). Randall Arendt. April 2004. 142pp.

525. E-Government. Jennifer Evans–Cowley and Maria Manta Conroy. May 2004. 41pp.

526. Codifying New Urbanism. Congress for the New Urbanism. May 2004. 97pp.

527. Street Graphics and the Law. Daniel Mandelker with Andrew Bertucci and William Ewald. August 2004. 133pp.

528. Too Big, Boring, or Ugly: Planning and Design Tools to Combat Monotony, the Too-big House, and Teardowns. Lane Kendig. December 2004. 103pp.

529/530. Planning for Wildfires. James Schwab and Stuart Meck. February 2005. 126pp.

531. Planning for the Unexpected: Land-Use Development and Risk. Laurie Johnson, Laura Dwelley Samant, and Suzanne Frew. February 2005. 59pp.

532. Parking Cash Out. Donald C. Shoup. March 2005. 119pp.

533/534. Landslide Hazards and Planning. James C. Schwab, Paula L. Gori, and Sanjay Jeer, Project Editors. September 2005. 209pp.

535. The Four Supreme Court Land-Use Decisions of 2005: Separating Fact from Fiction. August 2005. 193pp.

536. Placemaking on a Budget: Improving Small Towns, Neighborhoods, and Downtowns Without Spending a Lot of Money. Al Zelinka and Susan Jackson Harden. December 2005. 133pp.

537. Meeting the Big Box Challenge: Planning, Design, and Regulatory Strategies. Jennifer Evans–Cowley. March 2006. 69pp.

538. Project Rating/Recognition Programs for Supporting Smart Growth Forms of Development. Douglas R. Porter and Matthew R. Cuddy. May 2006. 51pp.

539/540. Integrating Planning and Public Health: Tools and Strategies To Create Healthy Places. Marya Morris, General Editor. August 2006. 144pp.

541. An Economic Development Toolbox: Strategies and Methods. Terry Moore, Stuart Meck, and James Ebenhoh. October 2006. 80pp.

542. Planning Issues for On-site and Decentralized Wastewater Treatment. Wayne M. Feiden and Eric S. Winkler. November 2006. 61pp.

543/544. Planning Active Communities. Marya Morris, General Editor. December 2006. 116pp.

545. Planned Unit Developments. Daniel R. Mandelker. March 2007. 140pp.

546/547. The Land Use/Transportation Connection. Terry Moore and Paul Thorsnes, with Bruce Appleyard. June 2007. 440pp.

548. Zoning as a Barrier to Multifamily Housing Development. Garrett Knaap, Stuart Meck, Terry Moore, and Robert Parker. July 2007. 80pp.

549/550. Fair and Healthy Land Use: Environmental Justice and Planning. Craig Anthony Arnold. October 2007. 168pp.

551. From Recreation to Re-creation: New Directions in Parks and Open Space System Planning. Megan Lewis, General Editor. January 2008. 132pp.

552. Great Places in America: Great Streets and Neighborhoods, 2007 Designees. April 2008. 84pp.

553. Planners and the Census: Census 2010, ACS, Factfinder, and Understanding Growth. Christopher Williamson. July 2008. 132pp.

554. A Planners Guide to Community and Regional Food Planning: Transforming Food Environments, Facilitating Healthy Eating. Samina Raja, Branden Born, and Jessica Kozlowski Russell. August 2008. 112pp.

555. Planning the Urban Forest: Ecology, Economy, and Community Development. James C. Schwab, General Editor. January 2009. 160pp.

556. Smart Codes: Model Land-Development Regulations. Marya Morris, General Editor. April 2009. 260pp.

557. Transportation Infrastructure: The Challenges of Rebuilding America. Marlon G. Boarnet, Editor. July 2009. 128pp.

558. Planning for a New Energy and Climate Future. Scott Shuford, Suzanne Rynne, and Jan Mueller. February 2010. 160pp.

559. Complete Streets: Best Policy and Implementation Practices. Barbara McCann and Suzanne Rynne, Editors. March 2010. 144pp.

560. Hazard Mitigation: Integrating Best Practices into Planning. James C. Schwab, Editor. May 2010. 152 pp.

561. Fiscal Impact Analysis: Methodologies for Planners. L. Carson Bise II. September 2010. 68pp.

562. Planners and Planes: Airports and Land-Use Compatibility. Susan M. Schalk, with Stephanie A. D. Ward. November 2010. 72pp.

563. Urban Agriculture: Growing Healthy, Sustainable Places. Kimberley Hodgson, Marcia Canton Campbell, and Martin Bailkey. January 2011. 148pp.